Expanding Horizons: Planning Space-Based Supply Chains for a Multi-Planetary Future

Dr. David Bigelow

DEDICATION

This book is dedicated to all the dreamers who envision human civilization expanding beyond Earth. The dreams are slowly morphing into reality. But there is still much work to be done.

CONTENTS

TABLE OF FIGURES

ACKNOWLEDGMENTS

I want to thank all the interesting and knowledgeable people I have met along my personal journey through the wonderful world of Aerospace. It started with my involvement on the Government oversight of a new potential launch company called SpaceX. This company proposed launch capabilities at price points that bordered on ludicrous. I am happy to have done my small part to open the door and allow SpaceX the opportunity to succeed or fail.

What I learned from that experience combined with the growing threats from adversaries drove me to continue working in this industry after retiring from the Army. This has led to multiple positions where I had perspectives of supply chains, project management, IT development, quality control, payload development and additive as well as traditional manufacturing of a variety of materials and systems. I have never grown bored in any of these positions and remain excited to see what the future holds as we boldly reach out across our solar system.

During all of this my family has been 100% supportive. My three sons are always an inspiration to watch as the grow and their minds mature. Their boundless curiosity and drive to succeed is infectious! Even more profound is my wife Karen who continues to support my endeavors while accomplishing great actions of her own. I would not be the man I am today without her support.

Introduction: Forging the Pathways Beyond Earth: The Dawn of Extraterrestrial Supply Chains

Imagine waking up in a habitat on the lunar surface. Outside your window, the Earth hangs in the black sky, a blue marble suspended in darkness. As you prepare for your day, you drink water extracted from lunar ice, eat food grown in hydroponic gardens fertilized with recycled waste, and put on a spacesuit partially manufactured from metals mined right here on the Moon. None of these essentials arrived on the last rocket from Earth—they are all products of one of humanity's greatest logistical achievements: the creation of supply chains beyond our home planet.

This isn't science fiction. The first elements of such a system are already being designed, and within decades, we may see humans living and working on the Moon, Mars, and perhaps even mining asteroids—not as visitors on short expeditions, but as residents of permanent settlements. The success of these bold ambitions hinges on a challenge more fundamental than rocket science: how will we move, transform, and utilize resources in the harsh environment of space?

This book delves into the intricate and multifaceted challenge of creating these extraterrestrial supply networks. It posits that the future of space exploration, colonization, and resource utilization is inextricably linked to our ability to forge reliable pathways for the flow of essential goods, resources, and personnel across the vast distances of space. Just as terrestrial civilizations have historically been shaped by their trade routes and logistical capabilities, our expansion into the cosmos will be defined by the effectiveness and ingenuity of our space-based supply chains.

Our journey begins in Chapter One lays the foundational principles of supply chain management and applies them to the unique challenges of the space environment. It establishes the core concepts of sourcing, transportation, storage, and distribution in a context where gravity, vacuum, radiation, and immense distances introduce complexities far beyond those encountered on Earth. Chapter Two builds upon this foundation, exploring the initial forays into space logistics and the existing supply lines that support our current orbital infrastructure, setting the stage for the more ambitious endeavors to come.

With this groundwork laid, Chapter Three turns our attention to the Moon, our closest celestial neighbor. It argues for the Moon's crucial role as the logical first step in building a truly space-based supply chain. Its relative proximity, coupled with the presence of valuable resources like water ice and regolith, and its lower gravity well, make it an ideal location for establishing initial extraterrestrial supply nodes. We explore the specific supply chain needs of a lunar base, the immense potential of In-Situ Resource Utilization (ISRU) to create a degree of self-sufficiency, the requirements for a lunar orbit infrastructure, and even the long-term possibilities of the Moon supporting activities in Earth orbit.

Building upon the lessons learned and technologies developed on the Moon, Chapter Four tackles the far more complex challenge of establishing a supply chain on Mars. The greater distance, similarly harsh environment, and longer mission durations associated with the Red Planet necessitate a significant leap in logistical sophistication and technological innovation. This chapter details the intricate supply chain needs of a Martian base, the critical role of advanced ISRU utilizing Martian atmospheric CO_2 and subsurface water ice, the potential for manufacturing and fabrication on Mars, and the orbital infrastructure required to support surface operations. We even touch upon the long-term and highly ambitious vision of returning precious Martian resources to Earth.

Chapter Five ventures into the vast expanse of the asteroid belt, highlighting its immense potential as a virtually untapped reservoir of metals, water ice, and volatile compounds. This chapter explores the unique

4

supply chain needs of operating in such a distant and dispersed environment, envisioning fleets of autonomous robotic miners and in-space processing facilities. We explore the potential outputs of an asteroid belt supply chain, including metals for space construction, water and propellant for deep space missions, and the long-term prospect of returning rare materials to Earth. Crucially, this chapter also address the significant challenges and ethical considerations associated with utilizing the resources of the asteroid belt, including the need for permanent supply stations to facilitate operations across this immense region.

Chapter Six delves deeper into the technology and infrastructure advances that exist and need to be created to support the grand plans. Advanced propulsion systems are critical to shortening transportation times from months to days. In-space manufacturing development will change these locations from Earth dependent to independent and eventually suppliers on their right. The creation of advanced life support, energy production, space ports and more will open to doors to long term habitation.

Chapter Seven reviews the strategic considerations for supporting solar system wide supply chains. This covers peripheral but important topics such as legalities and politics. It also posits potential economic models and business cases that will support the decisions to invest and more importantly when.

Throughout this book, the central thread remains the critical role of supply chains in enabling our expansion into space. Without the ability to reliably and efficiently move resources and personnel to where they are needed, the grand visions of lunar settlements, Martian colonies, and the utilization of asteroid resources will remain just that – visions. By meticulously examining the specific needs, challenges, and opportunities associated with establishing supply chains at each of these key locations, this book aims to illuminate the path towards a future where humanity is not confined to a single planet, but has forged the pathways beyond Earth, creating a truly interplanetary civilization sustained by the ingenuity and resilience of its extraterrestrial supply networks.

PART I: Foundations and Current Reality

Part 1 establishes the historical and contemporary context for space-based supply chains. I propose to you that humanity's journey into space, initially driven by Earth bound exploration and scientific curiosity, is now evolving towards a phase where sustained presence and economic activity beyond Earth orbit become increasingly vital. This section charts the progression from passive observation of space to active engagement, culminating in the recognition of the inherent limitations of an Earth-centric approach and the burgeoning need for in-space resourcefulness.

Figure 1. 1594 double hemisphere world map by Petrus Plancius. Copied from Wikipedia commons on April 9th, 2025.

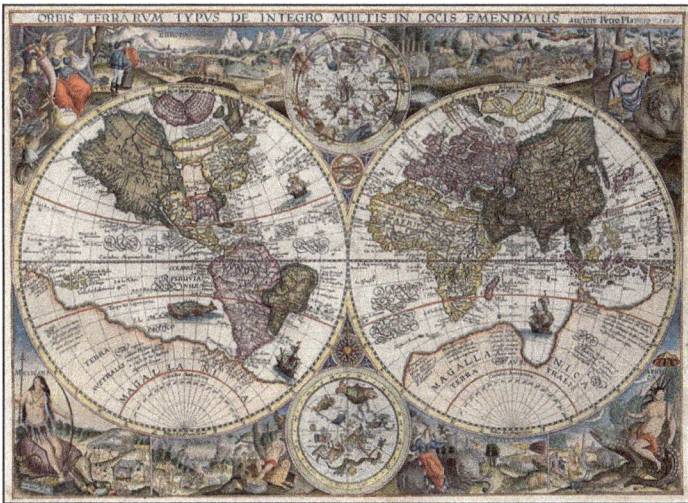

Note. Equal importance of having both geographic features and constellations for navigation.

Chapter One: From Celestial Observation to the Imperative for Space-Based Supply Chains

Put yourself in the sandals our of ancestors. You wake up before the sunrise and after the camp fires have all burned down. The night's sky is a dramatic vision that dominates your imagination. A vast swath of what appears to be tiny campfires burn all night, every night. Early humans had this dramatic scene and mysterious continuum to contend with. This scene captured their imagination and through careful study provided useful information. Our ancestors were profoundly impacted by both the warm light of the sun and the cold bright stars.

Humanity has a long-standing relationship with space, starting from pre-historic times. Seemingly passive observation of celestial phenomena has profoundly impacted human civilization, subtly setting the stage for our eventual deeper engagement with space. Celestial events directly affected early human societies, long before space travel was even dreamed about. The combination of information and solar energy is the first example of how humans have always relied upon the galactic supply chain for our survival and advancement.

Space has supplied us with numerous influences on Earthly Life. Early cultures (e.g., Egyptians, Babylonians, Mayans, Greeks) observed the movements of the Sun, Moon, planets, and stars. These observations led to the development of astronomy for timekeeping, navigation, and agricultural planning. Being able to understand and predict the changing seasons profoundly impact agricultural planning and harvests. This became essentially the earliest form of village resource management and "supply chain" reliance on solar energy. Coastal communities mastered understanding of the tides to survive from the bounty of the sea. Although our ancestors didn't know it, the Moon's gravitational pull and tides influenced fishing, navigation, and coastal settlement patterns. The ability to gain access to essentials such as food, water and shelter determined the fate of early tribes. Switching from hunter gathering to agrarian laid the foundation for larger societies. These societies were able to travel farther and trade with others creating new supply chains beyond the nature to human connection.

Navigation and exploration replied on the stars for land and seafaring navigation, enabling exploration and trade routes (Penprase, 2023). These became humanity's early global "supply chains" for trading of basic goods. Remarkably small wooden ships carried humans across the vast expanses of the oceans with confidence. This allowed humans to colonize new parts of the world and maintain connections to their points of origin. Figure 1 displays this importance of the celestial elements through a navigational map that contains both geographic features and key constellations.

Early philosophical and scientific inquiry focused on early cosmological models and the development of astronomy as a science, demonstrating humanity's innate curiosity about the universe that provided so much important information. These models inspired religions and proved culturally significant in every early civilization. Thus our solar system supplied us with far more than just physical nourishment.

Why Humans Created Constellations

Ancient cultures across the globe observed the seemingly random scattering of stars and, through pattern recognition and storytelling, grouped them into constellations. There were several key reasons for this (Saeed & Azeema, 2025). Constellations provided a framework for memorizing the vast number of visible stars. It's much easier to remember a few dozen distinct patterns than thousands of individual points of light. These patterns often became associated myths and legends, further aiding recall and cultural transmission of astronomical knowledge.

The Role of Stars in Ancient Navigation

For millennia, before the advent of GPS and sophisticated instruments, humans relied on the predictable patterns of the night sky for orientation and navigation (Penprase, 2023). The stars, appearing fixed in their relative positions over human timescales, provided a consistent and reliable celestial compass. This became crucial information when navigating beyond the edge of local areas of easily memorized terrain. This navigation accuracy, supplied by the stars, opened up the world to human exploration. The impacts were so profound that they became incorporated into every culture's intellectual and religious development.

Timekeeping was another invention derived from the study of constellations (Saeed & Azeema, 2025). The apparent movement of constellations across the night sky throughout the year allowed people to track the seasons, which was crucial for agricultural planning (planting, harvesting) and religious festivals. The rising and setting of specific constellations marked important times of the year. This led to the creation of more advanced time measuring devices and greater organizing of human activity.

Navigation on Land Using Stars

On land, where distinct landmarks are far apart (such as deserts, steppes or vast plains), the stars offered a reliable way to maintain a course. In the Northern Hemisphere, Polaris, also known as the North Star, is located very close to the North Celestial Pole. This means that as the Earth rotates, all other stars appear to revolve around Polaris. Therefore, Polaris remains almost stationary in the northern sky and always points north. Travelers could find Polaris and then orient themselves accordingly.

Other Constellations and Bright Stars were necessary to provide navigation data. When navigating in the southern hemisphere, there isn't a single bright star directly at the South Celestial Pole (as indicated by the search result). Additional constellations were commonly used for general orientation and direction, especially when Polaris or the Crux was obscured. For example, the position of Orion's belt could give a general sense of east and west.

Navigation at Sea Using Stars

For seafarers, navigating vast and featureless seas and oceans, the stars were absolutely essential (Penprase, 2023). **Latitude determination was and remains critical to sea faring.** The altitude (angle above the horizon) of Polaris is approximately equal to the observer's latitude in the Northern Hemisphere. By measuring the angle of Polaris above the horizon using instruments like an astrolabe or quadrant, sailors could determine their position north or south of the equator.

Sailors also use the rising and setting points of specific stars and constellations on the horizon to determine cardinal directions. For instance, certain stars consistently rise in the east and set in the west. When navigating in the southern hemisphere, the Crux is the again the constellation of choice. The longer axis of Crux points roughly towards the South Celestial Pole. By finding Crux, sailors could determine south and then orient themselves accordingly. Other bright stars and constellations like Canopus and the Magellanic Clouds were also important for navigation in the southern skies.

Figure 2. Travel routes and constellations mirroring each other.

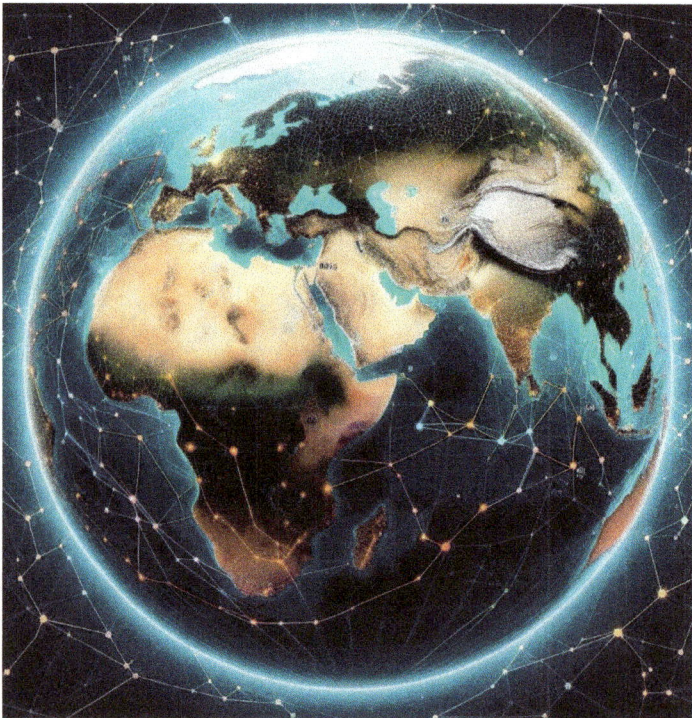

Timekeeping and longitude were also developed as key tools for accurate navigation. While not as straightforward as latitude, the movement of stars was also crucial for celestial navigation to estimate longitude, especially after the invention of accurate marine chronometers in the 18th century. Before that, estimations of longitude at sea were much more challenging and relied on dead reckoning combined with celestial observations.

The Sun: The Ultimate Driver of Life and Earth's Systems

The Sun is the primary source of energy for our planet, and its influence permeates nearly every aspect of Earth's environment and life. This supply of energy supports all forms of life on spaceship Earth. It's support continues to grow as our civilizations endeavor to refine the efficiencies of collecting this energy and harnessing it for our uses. **Agricultural development drove the growth of humanity. We gained more control over a consistent and nutritious food source that supported larger populations. These larger populations then allowed for more specialized skill development which led to a far more advanced civilization that we enjoy today.**

The most fundamental impact of the Sun on agriculture is photosynthesis. Plants use sunlight, water, and carbon dioxide to produce their own food (sugars) and release oxygen. This process forms the base of almost all terrestrial food chains. Without sunlight, agriculture as we know it would be impossible. The tilt of the Earth's axis and its orbit around the Sun cause variations in the amount of sunlight received at different latitudes throughout the year. This results in distinct seasons, which dictate the growing seasons for various crops. Farmers rely on the predictable cycle of seasons, driven by the Sun's apparent movement in the sky, to plan planting and harvesting. Sunlight not only provides energy but also influences various stages of plant development, including germination, flowering, and fruiting. The duration and intensity of sunlight (photoperiod) are critical factors for many plant species. Historically, and even today, solar energy is used for drying crops like grains, fruits, and meats, a crucial method of food preservation. Preserved foods supported longer trips and deliveries of foods thousands of miles from their point of origin.

Weather Patterns are also profoundly impacted by our sun. Uneven heating of the Earth's surface by the Sun drives global atmospheric circulation patterns. Warmer air at the equator rises and moves towards the poles, while colder air from the poles moves towards the equator. This differential heating creates pressure differences that result in winds, which are a fundamental component of weather. Solar energy causes water to evaporate from oceans, lakes, and land surfaces, forming water vapor in the atmosphere. This water vapor eventually condenses to form clouds and falls back to Earth as precipitation (rain, snow, hail), which is essential for agriculture and freshwater resources. The energy from the Sun fuels many weather phenomena, including thunderstorms, hurricanes (or typhoons), and cyclones. These powerful storms are driven by the transfer of heat and moisture in the atmosphere, ultimately originating from solar energy.

The Sun's influence extends to the very foundations of our planet's environment and, consequently, all terrestrial supply chains. Regarding **climate,** the Sun is the primary determinant of **Earth's average temperature,** with the balance of incoming solar radiation and outgoing terrestrial radiation dictating our planet's heat. This temperature range is crucial for the functioning of all natural and human-engineered systems that underpin our supply chains, from agriculture to transportation infrastructure. **Long-term climate**

variations, driven by subtle shifts in Earth's orbit around the Sun and changes in solar output, can have profound and lasting impacts on these supply chains. The Milankovitch cycles, for instance, which contribute to glacial and interglacial periods, have historically reshaped landscapes, altered agricultural zones, and influenced the availability of natural resources, directly affecting the reliability and structure of supply chains over extended periods.

Beyond climate, the Sun has **other impacts** vital to life and, therefore, our supply systems. **Solar radiation,** while the engine for photosynthesis and thus all food production, also presents challenges. Ultraviolet (UV) radiation, while necessary for vitamin D synthesis in humans (supporting a healthy workforce for supply chains), can also be harmful in excessive doses, impacting human health and the ability to work outdoors, a critical aspect of many Earth-based supply chains. The **day-night cycle,** a direct consequence of Earth's rotation relative to the Sun, dictates the rhythm of human activity and significantly influences the operational schedules of supply chains across the globe, from farming and manufacturing to transportation and logistics. Understanding these fundamental solar influences on Earth's environment is not only crucial for managing our current terrestrial supply chains but will also be essential as we venture into space and begin to establish supply chains beyond Earth, where we will need to contend with vastly different solar radiation levels and the absence of a natural day-night cycle in many environments.

While the Sun serves as Earth's primary energy source, the Moon exerts a more subtle yet still significant influence across various

Earth systems, with implications for both terrestrial and future space-based supply chains. In **agriculture,** the belief in the influence of **lunar cycles on planting** has persisted in many traditional practices, suggesting a connection between lunar phases and plant growth. Although modern science has largely not confirmed significant direct effects on crop yields, the cultural significance of these beliefs highlights humanity's long-standing awareness of celestial rhythms and their potential link to food production, a foundational supply chain. Furthermore, the Moon's role in creating tides could have subtle indirect effects on soil moisture in coastal agricultural regions. The Moon's gravitational pull is the primary driver of Earth's **tides, impacting weather patterns.** While the direct impact on atmospheric weather is still in early stages of research and considered less significant than solar influences, the powerful forces shaping our oceans have clear implications for coastal communities and maritime transportation, a critical component of global supply chains.

Perhaps one of the most crucial long-term impacts of the Moon is on **climate**, specifically the **stabilization of Earth's axial tilt.** This stability ensures relatively predictable seasons, which are fundamental for reliable agricultural cycles and the consistent production of food that underpins human societies and their supply chains. Without this stabilizing influence, more extreme and unpredictable climate swings could severely disrupt terrestrial food production. A well-known impact of the Moon is, of course, the **tides.** These have profound effects on coastal ecosystems, influencing the distribution and abundance of marine life, which directly impacts fishing industries – a vital part of the global food supply chain. Tides also play a crucial role in shipping and navigation, affecting the efficiency and cost of transporting goods across oceans. The growing potential for harnessing tidal energy represents another form of supply chain, providing a new renewable energy source. **Lunar light,** though not an energy source in itself, can influence the behavior of nocturnal animals, potentially affecting ecosystems that indirectly support supply chains, such as pollination by night-flying insects. As we look towards establishing supply chains in space, understanding the subtle but persistent influences of celestial bodies like the Moon on Earth's systems provides valuable context for how we might interact with and utilize resources in other parts of our solar system.

Interplay Between the Sun and Moon

It's important to recognize that the Sun and Moon often work in concert. For example, the Sun's heating drives evaporation, which contributes to the water that the Moon's gravity pulls into tides along coastlines. The combination of solar and lunar gravitational forces determines the magnitude of spring tides (when the Sun, Earth, and Moon are aligned) and neap tides (when they form a right angle).

The Sun and Moon are fundamental to life and the Earth's systems. The Sun provides the energy that drives photosynthesis, weather patterns, and maintains Earth's temperature, while the Moon, through its gravitational influence, stabilizes our planet's climate and creates the tides that shape our coastlines. Understanding these celestial influences has been crucial for human survival and development, particularly in areas like agriculture and navigation, and continues to be a vital area of scientific study. This knowledge will prove critical to planning successful off world supply chains around celestial bodies with different solar impacts and multiple lunar bodies such as Mars' Phobos and Deimos moons.

The Dawn of the Space Age and a New Paradigm

The dawn of the Space Age marked a pivotal shift in humanity's relationship with the cosmos, moving from an era of passive observation to one of active intervention. The launch of Sputnik in 1957, a seemingly simple artificial satellite, represented a monumental technological leap, showcasing the rapid advancements in rocketry and propulsion that finally made space travel a tangible reality. This breakthrough not only demonstrated a new capability to reach beyond Earth's atmosphere but also laid the foundation for transporting goods and eventually establishing supply lines into orbit. The initial forays into space were largely driven by scientific curiosity – a desire to understand the space environment, explore the solar system, and push the boundaries of human knowledge – and the intense geopolitical competition of the Cold War, where demonstrating technological superiority in space became a matter of national prestige. While these early missions weren't explicitly focused on supply chain development, they provided invaluable data and experience in operating in the harsh conditions of space, knowledge that would later prove critical for designing and implementing sustainable space-based logistics.

Following these initial exploratory phases came the first practical applications of space technology, most notably with the advent of communication and Earth observation satellites. These satellites revolutionized global communication, enabling near-instantaneous transmission of information across vast distances, effectively creating a new "supply chain" of data that underpinned emerging globalized economies. Similarly, Earth observation satellites provided unprecedented capabilities for weather forecasting, resource monitoring, environmental studies, and even military reconnaissance, establishing another vital information supply chain from space to Earth. These early applications, while not involving the movement of physical goods *to* space in large quantities, represent the nascent forms of space-based services and the first tangible economic and societal benefits derived from our ventures beyond Earth, setting the stage for the more complex and ambitious space-based supply chains envisioned for the future.

The Rise of Commercial Space and Expanding Applications

The latter half of the 20th and the early 21st centuries witnessed a significant shift in the landscape of space activities with the **rise of commercial space and an explosion of diverse applications.** Decades of post WWII space era were dominated by government-led exploration and military endeavors. The space sector began to see commercialization in the 1980s, most notably in the **satellite industry**. The bureaucratic transfer from Government only launches to licensed commercial launches took decades culminating in the first US licensed launch in March 1989 by Space Service, Inc. Others companies slowly emerged to capitalize on the unique vantage point of orbit for **telecommunications,** enabling global connectivity for phone calls, data transfer, and the internet. Simultaneously, the **broadcasting** industry leveraged satellites to deliver television and radio signals across vast distances. The development of sophisticated **Earth observation** satellites provided invaluable data for weather forecasting, environmental monitoring, agricultural management, and disaster response – all of which have become integral to the efficient operation of numerous terrestrial supply chains. This increasing economic value and societal reliance on space-based infrastructure underscored the critical role of space in supporting and enhancing existing Earth-bound supply networks.

More recently, the space sector has been further revolutionized by the emergence of the **"New Space" era**, characterized by significant **private sector involvement**. Driven by entrepreneurial spirit, technological innovation, and a desire to democratize access to space, companies like SpaceX, Blue Origin, and Rocket Lab have developed reusable rockets and innovative launch technologies, leading to a notable reduction in launch costs. This decrease in the barrier to entry has fostered a more diverse range of space activities and opened up new possibilities for future space-based supply chains by making transportation to and from orbit more economically accessible.

This new era has also ushered in an **expanding landscape of space applications**, each with potential implications for both terrestrial and future space-based supply chains. **GPS and other Global Navigation Satellite Systems (GNSS)** have become ubiquitous, providing essential positioning and timing data that underpin the logistics and transportation networks of global supply chains, enabling precise tracking of goods and efficient route optimization on Earth (Weiss, 2021). This system also provides the vital timing element of all digital money transactions. Every time you send a digital payment or use an ATM the systems coordinate and verify the transactions using GNSS. The advent of **internet from space** constellations promises to bring broadband connectivity to underserved and remote areas, potentially improving communication and coordination for supply chain operations in geographically challenging locations. While still in its nascent stages, **space tourism** represents the beginnings of more routine human access to space, which could eventually play a role in the development and maintenance of in-space infrastructure for future supply chains. The **experimental field of space-based manufacturing and research**, conducted on platforms like the International Space Station and through private initiatives, hints at the potential for producing specialized goods in microgravity or conducting research that could lead to breakthroughs in materials science, pharmaceuticals, and other areas relevant to building and operating supply chains beyond Earth. These developments collectively demonstrate a growing integration of space-based capabilities into terrestrial systems and lay the groundwork for the eventual establishment of self-sustaining supply chains in the vast expanse beyond our planet.

Limitations of Earth-Centric Supply: Setting the Stage for Change

The current reliance on an Earth-centric supply model for all space activities faces fundamental limitations that will become increasingly pronounced as our ambitions in space grow. **Reiterating the constraints,** it's crucial to remember the immense costs associated with overcoming Earth's gravity well, the inherent risks involved in each launch, and the logistical inefficiencies of transporting all necessary resources from our planet. This current supply chain, stretching from terrestrial manufacturing and assembly to the launchpad and finally to orbit, is complex, expensive, and high risk. As the demand for space-based services continues to surge, driven by burgeoning sectors like satellite constellations, space tourism, and the nascent field of in-space manufacturing, this purely Earth-dependent model is proving increasingly insufficient to meet demand. The sheer volume of materials required for more ambitious endeavors, such as lunar bases, Martian missions, and asteroid mining, makes the prospect of launching everything from Earth not only economically prohibitive but also logistically daunting. Therefore, to truly unlock the potential of space for scientific discovery, resource utilization, and human expansion, a fundamental shift is required. This necessitates **introducing the need for space-based supply chains,** a paradigm where resources are sourced, processed, manufactured, and transported within space itself. This transition, which will be explored in the following chapters, is not merely an optimization but a crucial prerequisite for establishing a sustained and thriving presence beyond

Earth, creating new economic opportunities both in space and back on our home planet.

Chapter Two: Current Earth Orbit Supply Chains: Launch Dependence and Emerging In-Space Activities

The Current Paradigm: Earth-to-Orbit Launch as the Primary Supply Line

Humanity has long associated the celestial bodies with the power of gods. Mars, the Moon and Milky Way have all been given deity titles and assumed to have impacts on Earthly activities. In the case of the Moon, this is accurate with its influence on tidal conditions. The sun also has profound influences on human activities. Only recent human history have we ventured into orbit and created human manufactured objects that also impact our daily lives. There is no turning back from the extensive use of satellites but rather there is a growing hunger for more.

The story of space exploration and utilization, from its nascent beginnings to the sophisticated infrastructure we see today, is fundamentally intertwined with the capability to launch materials and personnel from Earth into orbit. For over six decades, the Earth-to-Orbit (ETO) launch has served as the almost exclusive artery for supplying everything needed for activities beyond our planet's atmosphere. This existing model, while remarkably successful in enabling a wide range of scientific, commercial, and national security endeavors, operates under a paradigm of complete dependence on terrestrial resources and infrastructure. Understanding this current supply line is crucial for appreciating its inherent limitations and the growing imperative for developing alternative, space-based solutions.

The process of acquiring the necessary supplies to build sophisticated space vehicles and launch vehicles is fraught with risks that can significantly impact project timelines and budgets. Long lead items are a prime example of this vulnerability. These components often have extended procurement times driven by a confluence of factors. One example is Beryllium, a lightweight and stiff metal with specialized applications in aerospace but processed in very small quantities that economically support few processors. Another case is the large, high-precision optical mirrors required for space telescopes like the James Webb Space Telescope. The process of fabricating these mirrors, often involving specialized glass materials, intricate polishing techniques, and cryogenic testing, can take many years and involves a limited number of facilities worldwide with the necessary expertise. Similarly, certain specialized alloys like Inconel or Haynes, crucial for the extreme temperature and pressure environments within rocket engines, require specific and lengthy manufacturing processes, often involving limited production capacity and strict quality control. These long lead times can create bottlenecks in the overall vehicle production schedule, delaying critical milestones in the development of space-based supply chain infrastructure.

High demand items also pose a significant risk. Computer processing boards are a well-known example. The aerospace industry often relies on specific grades of radiation-hardened electronics, such as microprocessors, memory chips, and Field-Programmable Gate Arrays

(FPGAs). These components are designed to withstand the harsh radiation environment of space, and their production involves specialized manufacturing processes and rigorous testing. The demand for these components not only comes from the aerospace sector but also from military applications and other high-reliability industries, leading to potential shortages and extended lead times, especially when global events or technological shifts impact the supply of semiconductor manufacturing capacity or the demand priorities for existing production. This scarcity can drive up costs and delay the integration of critical avionics and control systems in both launch and space vehicles essential for establishing off-world supply lines.

Reliance on sole source producers of low rate production raw materials introduces another layer of risk. The example of rayon, potentially used in certain solid rocket motor components, illustrates this point. If the single company producing a specific, qualified material faces production issues due to technical problems, natural disasters, or even economic factors, the entire supply chain can be severely disrupted. Consider also specialized sealants and adhesives qualified for the extreme temperature variations and vacuum of space. If only one or a very limited number of companies possess the proprietary formulations and manufacturing capabilities for these critical items, the aerospace industry becomes highly vulnerable to any disruption in their supply. This lack of redundancy can halt production lines and significantly impact the timelines for building vehicles needed to transport resources and personnel for off-world supply chains.

Finally, shifting ownerships due to buyouts can create instability and uncertainty within the aerospace supply chain. When a key supplier is acquired by another company, several risks can emerge. The new ownership might have different priorities, potentially leading to a discontinuation of previously supplied materials or components if they no longer align with the acquiring company's strategic direction. A buyout could also result in changes in management, personnel, or established quality control processes, potentially impacting the reliability, pricing and consistency of the supplied parts. Furthermore, the acquiring company might be a competitor in the aerospace sector, potentially leading to prioritization of their own internal needs over external contracts or even the strategic withholding of critical components. These disruptions in established supplier relationships can force aerospace manufacturers to seek new sources, requiring time-consuming and costly re-qualification processes, ultimately delaying the production of vehicles vital for building space-based supply chains.

The existing space supply chains are almost entirely Earth-based, meaning that every kilogram of material, every component, and every astronaut destined for orbit or beyond originates from our planet. This fundamental constraint imposes significant limitations and costs due to the energy required to overcome Earth's gravity and the atmospheric drag encountered during launch. Currently, the most prominent examples of ongoing supply chains to orbit are the missions dedicated to resupplying orbital stations. The International Space Station (ISS), a collaborative project involving multiple international partners, receives regular resupply missions from various providers.

SpaceX's Dragon spacecraft and Northrop Grumman's Cygnus spacecraft, launched on their respective Falcon 9 and Antares rockets, routinely deliver essential cargo to the ISS under commercial contracts with NASA. These missions transport a wide array of supplies, including food, water, breathable air, scientific equipment, spare parts for maintaining the station's systems, and even new experiments. Similarly, China's Tiangong space station is also supported by regular resupply missions using its Tianzhou cargo spacecraft. These resupply efforts represent the current extent of sustained space-based logistics, primarily focused on maintaining human presence and scientific research in Low Earth Orbit (LEO).

Beyond these resupply missions, the vast majority of satellites launched into various orbits are designed with a limited operational lifetime. This lifespan is typically dictated by the depletion of onboard consumables, most notably propellant used for station-keeping and attitude control, but can also be affected by the degradation of electronic components due to the harsh radiation environment of space and the occasional impact from micrometeoroids or orbital debris. Once a satellite reaches its end of life, current protocols dictate its disposal. For satellites in lower orbits, the common practice is to perform a deorbit maneuver, using any remaining propellant to lower its altitude so that it will eventually enter the Earth's atmosphere and burn up during reentry. Satellites operating in or near Geosynchronous Earth Orbit (GEO), where deorbiting is a much more energy-intensive process, are often boosted to a higher, less congested orbit known as a graveyard orbit or super synchronous orbit, typically a few hundred kilometers above GEO.

These satellites parked in a geosynchronous orbit represent a potential untapped resource within a future space-based supply chain. While currently considered space debris, these retired satellites have the potential to be resupplied with propellant and have their operational lifespan extended, or even have their capabilities upgraded through the addition of new modules via on-orbit servicing missions (as hinted at in Chapter Six). Emerging technologies and companies are actively developing the capabilities for robotic spacecraft to rendezvous with, inspect, repair, refuel, and even upgrade existing satellites in orbit. Furthermore, these end-of-life satellites could themselves be used as a resource through salvage. They contain significant amounts of valuable materials, including aluminum, titanium, and various electronic components. In the future, robotic spacecraft could be tasked with dismantling these defunct satellites and harvesting their usable parts and materials, which could then be repurposed for in-space manufacturing (Chapter Six) of new satellites, components for space stations, or even construction materials for lunar or Martian habitats. This concept aligns with the principles of a circular economy in space, reducing the need to launch all materials from Earth and mitigating the growing problem of space debris. The ability to repurpose and reuse existing orbital assets will be a crucial element in building a sustainable and cost-effective space-based supply chain.

The Earth-based aerospace supply chain (Figure 3) is a complex ecosystem with numerous inherent risks. The examples of long lead items like specialized alloys and optical mirrors, high-demand radiation-hardened electronics, sole-source producers of critical materials like specialized sealants, and the instability caused by shifting ownership highlight the vulnerabilities that must be actively and diligently managed to ensure the timely and efficient production of the space and launch vehicles that will be the backbone of our future off-world supply chains. Proactive risk assessment, diversification of suppliers where possible, strategic stockpiling of critical items, and the development of robust contingency plans are essential to mitigate these challenges and keep our ambitions for space exploration and resource utilization on track.

The current Earth to orbit supply chain doesn't just include the materials and services to produce the end products. It begins with the conception, design, and manufacturing of payloads on Earth. These payloads can range from small CubeSats weighing just a few kilograms to massive telecommunications satellites or components for the International Space Station (ISS) weighing several tons. The terrestrial supply chain involved in creating these payloads is complex and mirrors many industries on Earth, encompassing the sourcing of raw materials, the fabrication of intricate components, rigorous testing and quality control procedures, and the final assembly of the spacecraft. Once the space vehicle is complete, it is transported to a launch facility, often located strategically near coastlines to allow for over-ocean launches and minimize risks to populated areas.

Figure 3. Author's aerospace supply chain pyramid.

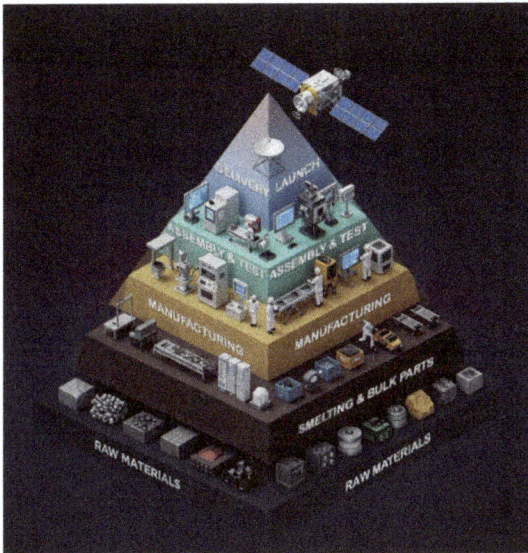

The launch industry itself is a diverse ecosystem comprised of various players, each with their own technological approaches, target markets, and operational philosophies. Historically, government space agencies like NASA (National Aeronautics and Space Administration) in the United States, Roscosmos in Russia, the European Space Agency (ESA), the China National Space Administration (CNSA), the Japan Aerospace Exploration Agency (JAXA), and [1] the Indian Space Research Organization (ISRO) were the primary entities responsible for developing and operating launch vehicles. Their initial focus was largely on scientific exploration, national prestige, and military applications. These agencies often developed their own launch capabilities or contracted with domestic aerospace companies.

However, the landscape of the launch industry has undergone a significant transformation in recent decades with the rise of commercial launch providers. Companies like SpaceX, Blue Origin, United Launch Alliance (ULA), Arianespace, and Rocket Lab have emerged as key players, offering a range of launch services to both government and commercial customers. SpaceX, for instance, has revolutionized the industry with its partially reusable Falcon family of rockets and its ambitious Starship program, aiming for significantly lower launch costs and the capability for large-scale transport to deep space. Blue Origin is developing the New Shepard for suborbital space tourism and the larger New Glenn for orbital missions. ULA, a joint venture between Lockheed Martin and Boeing, has a long history of providing reliable launch services for critical government payloads on the Atlas and Delta and this remains their primary customer with their new Vulcan rocket. Arianespace offers a suite of launch vehicles, including the Ariane and Vega, catering to a diverse range of payload sizes. Smaller companies like Rocket Lab focus on providing dedicated launch services for small satellites, a rapidly growing segment of the market.

The types of payloads that are launched into Earth orbit are equally varied, each with its own specific supply chain requirements and operational goals. **Satellites** represent the most mature and prevalent type of payload. These can be broadly categorized by their function.

- **Communication Satellites:** Facilitating global telecommunications, internet access, and broadcasting services. The supply chain for these involves complex electronics, high-power systems, and precise antenna technology.
- **Earth Observation Satellites:** Providing critical data for weather forecasting, climate monitoring, disaster response, and resource management. These require sophisticated sensors and imaging equipment.

- **Navigation Satellites:** Enabling global positioning systems (GPS), GLONASS, Galileo, and BeiDou, essential for countless applications on Earth. Their supply chain emphasizes accuracy and reliability.

- **Scientific Research Satellites:** Conducting experiments in microgravity, observing astronomical phenomena, and studying Earth's environment. These often involve highly specialized and custom-built instruments.

- **Military and National Security Satellites:** Providing reconnaissance, surveillance, and secure communication capabilities. Their supply chain is often shrouded in secrecy and involves stringent security protocols.

Beyond satellites, **human spaceflight missions** represent another growing demand on the Earth orbit supply chain. Missions to the ISS, for example, require the launch of crew capsules like the Russian Soyuz and the SpaceX Dragon, as well as cargo resupply vehicles such as the Russian Progress, the Northrop Grumman Cygnus and Dragon, and potentially future commercial vehicles. The supply chain for human spaceflight is incredibly complex, encompassing not only the spacecraft themselves but also life support systems (providing breathable air, water, and temperature regulation), food, medical supplies, scientific equipment for experiments conducted on board, and even personal items for the astronauts.

Space probes and telescopes destined for deep space or specific Earth orbits also rely entirely on Earth launch. These missions, such as the James Webb Space Telescope or interplanetary explorers like the Mars rovers, often involve highly specialized and meticulously crafted instruments and spacecraft, with unique supply chains tailored to their specific mission requirements and harsh operating environments.

Finally, the **International Space Station (ISS)** itself is a testament to the capabilities of the Earth orbit supply chain. This massive orbital laboratory was assembled piece by piece through numerous launches over more than two decades. The supply chain for the ISS continues to be active, with regular resupply missions bringing essential consumables, spare parts, and new scientific equipment to the orbiting outpost. New 3D printing experiments are now being done on the ISS. The first products were plastic materials, but a metal printer successfully operated in January 2024. These experimental prints mark the first steps in orbital manufacturing.

The cost of launching these diverse payloads into Earth orbit is a significant factor that shapes the entire space industry and its associated supply chains. Several key factors drive these costs:

- **Propellant Costs:** Rockets require vast amounts of propellant to achieve the necessary velocity to reach orbit. The type of propellant used (e.g., kerosene, liquid hydrogen, liquid oxygen) and its performance characteristics directly impact the amount needed and the associated expenses for production, storage, and handling.
- **Manufacturing and Assembly:** Rockets are complex machines with thousands of intricate parts, and stringent quality requirements. The manufacturing and assembly processes require highly skilled labor, specialized facilities, and advanced materials, all contributing to the overall cost.

- **Infrastructure and Operations:** Launch facilities are expensive to build and maintain. The operational costs associated with a launch include personnel (engineers, technicians, safety officers), tracking stations, control centers, and the logistics of preparing and executing a launch.

- **Research and Development:** The development of new launch technologies and vehicles requires significant upfront investment in research, engineering, and testing. These up-front investment costs are normalized into the price of launch services.

- **Insurance and Regulatory Compliance:** The space industry is inherently risky, and launch providers typically carry significant insurance policies to cover potential failures. Compliance with various national and international regulations also adds to the overall cost.

- **Economies of Scale:** Like any industry, the cost per launch can be reduced by increasing the volume of launches and streamlining production processes. However, the relatively low launch cadence compared to other industries limits the potential for significant economies of scale. SpaceX is the only company currently capable of utilizing this option.

- **Reusable vs. Expendable Launch Systems:** The advent of reusable launch systems, pioneered by companies like SpaceX, has the potential to dramatically reduce launch costs by recovering and reusing expensive rocket stages. This shift towards reusability is a significant development in the Earth orbit supply chain.

Providing precise average costs for space launches is challenging due to the proprietary nature of contracts and the wide variability depending on numerous factors. However, there are general ranges and trends based on publicly available information and common knowledge within the space industry. Keep in mind these are broad estimations as of late 2024/early 2025, and costs can change.

Here's a breakdown by approximate space vehicle size and common final orbits:

Categorization by Space Vehicle Size (Approximate Payload Capacity to LEO):

- Small: Up to ~1,000 kg (1 metric ton)
- Medium: ~1,000 kg to ~20,000 kg (1 to 20 metric tons)
- Large: Over ~20,000 kg (20+ metric tons)

Average Cost Ranges (USD per kilogram of Payload):

ORBIT TYPE	SMALL LAUNCH VEHICLES	MEDIUM LAUNCH VEHICLES	LARGE LAUNCH VEHICLES
LOW EARTH ORBIT (LEO)	$10,000 - $30,000+	$3,000 - $15,000	$2,000 - $8,000
SUN-SYNCHRONOUS ORBIT (SSO)	$12,000 - $35,000+	$4,000 - $18,000	$2,500 - $9,000
GEOSTATIONARY TRANSFER ORBIT (GTO)	$15,000 - $40,000+	$5,000 - $25,000	$3,000 - $12,000

Key Observations and Influencing Factors

Small Launch Vehicles

These are typically used for deploying smaller satellites (e.g., CubeSats, Small Sats) or for dedicated launches of single, specialized payloads. The cost per kilogram tends to be higher due to the smaller payload capacity and the low rate of flights. The insurance risks can also be higher for less frequently flown vehicles. Examples include Rocket Lab's Electron, Virgin Orbit's Launcher One (though Virgin Orbit went bankrupt in 2023, its assets and technology may be revived), and various emerging small launch providers.

Medium Launch Vehicles

This is a broad category encompassing workhorse rockets used for a wide range of missions, including deploying larger constellations, scientific satellites, and cargo to the International Space Station (ISS). The cost per kilogram is generally lower than small launch vehicles due to economies of scale and more established technologies. Examples include SpaceX's Falcon 9 (both expendable and reusable versions), United Launch Alliance's (ULA) Atlas V and Vulcan Centaur, and Arianespace's Soyuz and Vega.

Large Launch Vehicles

These are used for very heavy payloads, such as large telecommunications satellites, major scientific missions (like the James Webb Space Telescope), and crewed missions. While the overall mission cost is very high, the cost per kilogram can be the lowest, especially for very large payloads. Examples include SpaceX's Falcon Heavy, ULA's Delta IV Heavy & Vulcan, and potentially future vehicles like NASA's Space Launch System (SLS) and Blue Origin's New Glenn.

Impact of Orbital Location on Space-Based Supply Chains

The choice of orbital location significantly impacts the energy requirements, accessibility, coverage, and overall cost-effectiveness of various components within a space-based supply chain. Each orbit offers unique advantages and disadvantages that must be carefully considered when designing logistical networks for lunar, Martian, and asteroid belt operations.

LEO (Low Earth Orbit): As the orbit closest to Earth, typically ranging from altitudes of around 160 to 2,000 kilometers, LEO is indeed the easiest orbit to reach, requiring the least amount of energy for launch and time to orbit. Consequently, it generally has the lowest cost per kilogram to place payloads in this region.

• Benefits for Supply Chains: Its proximity to Earth also allows for relatively low-latency communication, which is beneficial for controlling robotic operations or transferring data. LEO's relative accessibility makes it an ideal staging ground for the initial assembly of large spacecraft or orbital transfer vehicles destined for the Moon, Mars, or the asteroid belt (as discussed in Chapter Six). LEO could also serve as a location for initial refueling depots (Chapter Six), where propellant launched from Earth or potentially sourced from the Moon could be stored before being transferred to deep-space vehicles.

• Disadvantages for Supply Chains: The close proximity to Earth's gravity well and the high orbital velocity required to maintain altitude in LEO means that spacecraft experience significant atmospheric drag, necessitating frequent and costly reboosting maneuvers to prevent orbital decay. Furthermore, a single satellite in

LEO has a limited coverage area over the Earth's surface at any given time, requiring constellations of satellites for continuous global coverage, which can be complex and expensive to deploy and maintain. While useful for initial stages, LEO is not suitable for direct transfer to geostationary orbit or deep space without significant additional propulsion, adding complexity to the overall supply chain.

SSO (Sun-Synchronous Orbit): This specialized type of LEO is designed such that the satellite passes over any given point on the Earth's surface at the same local solar time. Achieving this requires a specific inclination, typically near-polar, which typically adds a bit to the cost compared to LEO.

•   Benefits for Supply Chains: The consistent lighting conditions in SSO are ideal for Earth observation satellites, which play a crucial role in monitoring terrestrial resources and environmental changes that might impact the overall space-based supply chain. The predictable overpass times allow for scheduled data collection, which can be valuable for tracking the movement of resources or monitoring the status of infrastructure on Earth. SSO can also be useful for certain types of scientific missions that require consistent solar illumination. Constant solar radiation also means access to that energy source which can translate to a reduced on board battery requirements.

•   Disadvantages for Supply Chains: Reaching the higher inclination of SSO requires more energy than reaching a standard LEO at a lower inclination, thus increasing launch costs. While often at higher altitudes within the LEO range, SSO spacecraft are still subject to atmospheric drag and may require occasional reboosting. Similar to other LEO orbits, SSO is not directly conducive to transfers to geostationary orbit or deep space without additional propulsion stages and maneuvers.

MEO (Medium Earth Orbit): Situated between LEO and GEO, MEO typically ranges in altitude from approximately 2,000 kilometers to just below geostationary orbit at 35,786 kilometers.

• Benefits for Supply Chains: Satellites in MEO have a longer visibility time over a specific point on Earth compared to those in LEO, requiring fewer satellites for broader coverage. The radiation exposure in MEO is also lower than during transit through the Van Allen radiation belts encountered when moving to or from GTO. MEO is notably used by navigation satellite constellations like GPS, Galileo, and GLONASS. This could provide crucial navigation and timing data for spacecraft and robotic assets operating throughout the inner solar system, including lunar and Martian surface operations (as mentioned in Chapter Six). It requires less energy to reach than GEO, making it a potentially more accessible orbit for certain types of communication relays supporting Earth based and off-world activities.

• Disadvantages for Supply Chains: Reaching MEO requires more energy and time than reaching LEO, leading to higher launch costs. The communication latency is greater than in LEO, which might be a factor for real-time control of certain operations (laser communications systems are the next technological step to try and reduce latency). For continuous global coverage for applications like navigation or communication, a constellation of satellites is still required, though typically fewer than in LEO.

GTO (Geostationary Transfer Orbit): GTO is a highly elliptical orbit with its perigee (closest point to Earth) in LEO and its apogee (farthest point) at geostationary altitude (approximately 35,786 kilometers). Reaching GTO requires significantly more energy and time than LEO, SSO or MEO, resulting in higher costs per kilogram. The satellite then uses its own onboard propulsion to perform an apogee kick maneuver to circularize its orbit at geostationary altitude and adjust its inclination to reach GEO.

•   Benefits for Supply Chains: GTO is the most efficient transfer orbit for placing satellites into Geostationary Earth Orbit (GEO). GEO is ideal for communication and weather satellites that need to remain over a fixed point on Earth, providing continuous coverage. While not directly a location for supply chain operations beyond Earth, GEO communication satellites are vital for maintaining contact with lunar and Martian missions, and will be equally critical for future robotic missions in the asteroid belt.

•   Disadvantages for Supply Chains: Reaching GTO demands substantial launch energy and thus incurs higher costs per kilogram. Satellites deployed in GTO require significant onboard propellant for the final maneuvers to reach GEO, which reduces the mass available for the rest of the payload. The transit through the Van Allen radiation belts in the elliptical GTO can also pose risks to sensitive electronics, requiring additional shielding. GTO itself is a temporary orbit and not a suitable long-term location for supply depots or manufacturing facilities supporting off-world supply chains. GEO positions do provide the stability GTO's lack, but are significantly far from Earth and require tremendous energy and time to reach. Thus they are not optimal for supporting interorbital supply or inter solar system supply.

Understanding the trade-offs associated with each orbital location is fundamental to designing an efficient and effective space-based supply chain architecture that can support humanity's Earthly needs and expansion into the cosmos. The optimal choice of orbit will depend on the specific requirements of each element within the logistical network, balancing factors such as cost, accessibility, coverage, communication latency, and the intended destination.

Limitations of Current Launch Capabilities

Despite the remarkable achievements enabled by Earth-to-orbit launch, the current model faces several fundamental limitations that hinder further expansion and the development of more ambitious space endeavors. These limitations directly impact the feasibility and cost-effectiveness of establishing robust supply chains in Earth orbit and beyond.

One of the most significant limitations is the **high cost of launch.** As previously discussed, the complex processes and vast resources required to send even a single kilogram of mass into orbit result in exorbitant costs compared to terrestrial transportation. These high costs permeate every aspect of space activities, from the size and complexity of satellites to the scope and duration of human spaceflight missions. For commercial ventures, high launch costs are a barrier to entry for smaller companies and can significantly impact the profitability of all space-based services. For scientific missions, budget constraints often force compromises in the size and capabilities of spacecraft and instruments. The high cost of launch essentially acts as a significant tax on any activity that requires putting mass into space, severely limiting the scale and scope of what is currently achievable.

Another critical limitation is the existence of **limited launch windows**. Due to the precise nature of orbital mechanics, the alignment of celestial bodies, and the specific requirements of different missions, launch opportunities are often constrained to specific periods of time. For example, reaching the ISS requires precise timing to rendezvous with the orbiting station. Interplanetary missions have even more restrictive launch windows that occur only at certain intervals when the target planet is in a favorable position relative to Earth. These limited windows can lead to significant delays if a launch is postponed due to technical issues or weather conditions, potentially impacting mission timelines, scientific research schedules, and the deployment of critical infrastructure.

Reliability issues also pose a significant challenge to the current Earth orbit supply chain. While launch vehicle reliability has improved considerably over time, the inherent complexity and the extreme forces involved in launching a rocket mean that failures, though infrequent, still occur. A launch failure can result in the complete loss of valuable payloads, leading to significant financial losses, mission delays, and reputational damage for the launch provider and the payload operator. The need for robust safety measures and rigorous testing protocols adds to the cost and complexity of the launch process and can be driven by insurance requirements. The potential for launch failures introduces an element of unpredictability into the supply chain, which can be particularly problematic for time-sensitive deployments or critical resupply missions.

Environmental concerns associated with rocket emissions are an increasingly important impact. The combustion of rocket propellants releases various substances into the atmosphere, including carbon dioxide, water vapor, nitrogen oxides, and soot. While the overall contribution of rocket launches to global emissions is currently relatively small compared to other industries, the growing frequency of launches and the potential impact on the ozone layer and climate change are raising concerns. There is increasing pressure on the launch industry to develop "green" propellants and more sustainable launch practices to mitigate its environmental footprint. This shift towards more environmentally friendly technologies could potentially impact the supply chains for propellants and the design of future launch vehicles. It can also become a strong political driver for generating off planet supply chains.

Finally, the **"tyranny of the rocket equation"** represents a fundamental physical limitation of current launch technology. This equation describes the exponential relationship between the amount of propellant required for a rocket to achieve a certain velocity change and the mass of the payload it can carry. In essence, a vast majority of a rocket's mass at launch must be propellant, leaving a relatively small fraction for the actual payload. This limitation makes it incredibly challenging and expensive to send large amounts of mass to high Earth orbits, the Moon, Mars, or beyond. It also highlights the inherent inefficiency of launching everything needed for space activities from Earth's deep gravity well. Overcoming the tyranny of the rocket equation is a key driver behind the exploration of in-space resource utilization and the development of more efficient propulsion technologies.

The SpaceX Factor and Reusability

SpaceX has significantly disrupted the launch market by pioneering reusable rocket technology, particularly with its Falcon 9 and Falcon Heavy vehicles. This impact is significant enough to deserve a brief description. Reusability drastically reduces the cost of subsequent launches as the most expensive components (the first stage booster) are recovered and refurbished. This is a primary reason why SpaceX often offers some of the lowest costs per kilogram in the medium and large launch vehicle categories. This model is now the new standard that all launch companies are striving to emulate.

Important Considerations for launch planning:

• Averaging: Actual costs can vary significantly based on the specific mission requirements, the chosen launch provider, the contract terms, and other factors.
• Dedicated vs. Rideshare: The cost per kilogram can be very different if you are the sole payload on a rocket (dedicated launch) versus sharing the launch with other satellites (rideshare). Rideshare options can significantly reduce costs for smaller payloads.

- Emerging Launch Providers: The landscape of space launch is constantly evolving, with new companies and technologies emerging. These new entrants may offer different pricing structures.

- Inflation and Market Dynamics: Launch costs can be influenced by inflation and the overall supply and demand in the launch market.

While precise figures are difficult to obtain, these ranges provide a general understanding of the average costs associated with space launches based on vehicle size and final orbit. The trend towards reusability is a significant factor driving down costs in the medium and large launch vehicle categories.

Existing Near Earth Activities and Initial Supply Chain Evolution

While the dominant paradigm for supplying Earth orbit remains Earth-to-orbit launch, several existing in-space activities are beginning to incorporate elements of what could be considered nascent space-based supply chains. These activities, though currently limited in scope and still heavily reliant on Earth for primary resources, offer glimpses into a future where a more self-sufficient space economy might be possible.

The **satellite deployment and operations** industry, as the most mature space-based activity, provides a useful lens through which to examine the current state of space-based "supply." While the manufacturing and initial launch of satellites are entirely dependent on Earth-based supply chains, the ongoing operations of these satellites in orbit involve a limited form of in-space resource management. For instance, satellites utilize solar panels to generate power, effectively harnessing an in-space resource (solar radiation). They also carry onboard propellant for attitude control and station keeping, which, in the current model, is entirely supplied from Earth at launch. The end-of-life disposal of satellites, either through deorbiting or moving them to graveyard orbits, can also be viewed as a form of orbital space management. However, the vast majority of the lifecycle of a satellite, from its creation to its eventual demise, is dictated by the initial supply from Earth.

The **International Space Station (ISS)** serves as a compelling case study in the complexities and limitations of the current Earth orbit supply chain. As a permanently crewed orbiting laboratory, the ISS requires a continuous and substantial flow of supplies from Earth to sustain its operations and crew. While the assembly of the ISS involved significant in-space activity, with astronauts and robotic arms piecing together modules launched separately, the station remains fundamentally dependent on Earth for consumables like food, water, oxygen, and propellant for periodic reboosts to maintain its orbit. Astronauts do perform in-space repairs and maintenance, effectively acting as an in-space "service and repair" element of a limited supply chain. However, major component failures or the need for significant upgrades still typically necessitate launching replacement parts from Earth. The ISS, while a marvel of engineering and international collaboration, highlights the challenges of creating a truly self-sufficient system in Earth orbit under the current supply paradigm.

The emergence of **in-space services,** though still in its early stages, represents a significant step towards developing more robust space-based supply chain elements. **Satellite servicing and refueling** are gaining traction as a viable industry. Companies are developing robotic spacecraft capable of docking with existing satellites to extend their lifespan through station keeping maneuvers, attitude control adjustments, and even potentially refueling them. This capability directly addresses the limitation of having to launch an entirely new, expensive satellite when an existing one runs out of fuel or experiences a minor malfunction. In-space servicing creates a demand for in-space fuel depots and the manufacturing of servicing components, potentially forming the building blocks of a future in-space logistics network.

In-Space Manufacturing (ISM) is another nascent field with the potential to revolutionize space-based supply chains (Kulu, 2022). Experiments conducted on the ISS have demonstrated the feasibility of manufacturing certain materials and components in the unique microgravity environment of space. 3D printing has been used to create tools and replacement parts on demand, reducing the reliance on Earth-based resupply for these items. Research is also underway in areas like bioprinting and the production of specialized materials with unique properties achievable only in space. While still in its experimental phase, ISM holds the promise of enabling the on-demand production of goods needed in space, potentially reducing the mass and cost associated with launching everything from Earth. Sourcing raw materials for ISM in space, however, remains a significant challenge that will need to be addressed through future in-space resource utilization efforts.

Finally, the growing problem of **space debris** has spurred the development of technologies and strategies for its removal. While not directly producing goods or services, space debris removal can be seen as a crucial element of maintaining a usable and safe orbital environment, which is essential for the long-term sustainability of all space activities, including future in-space supply chains. Both active removal concepts, involving spacecraft that capture and deorbit debris, and passive mitigation measures, aimed at preventing the creation of new debris, contribute to the health and longevity of the orbital "commons." This nascent effort highlights the need for in-space services that support the overall infrastructure and sustainability of Earth orbit.

Challenges and Bottlenecks in Earth to Orbit Supply

The current Earth-dependent model for supplying activities in Earth orbit faces numerous challenges and bottlenecks that will need to be overcome to enable a more robust and sustainable space economy. The **high cost and inherent risks associated with Earth launch** remain the primary limitations. This single point of dependence on terrestrial infrastructure and resources creates vulnerabilities and limits the scalability of space activities. From a purely supply chain leader perspective, Earth is the sole source provider for space based systems. The current supply chain is also environmentally concerning, with the increasing frequency of launches raising questions about the long-term impact on Earth's atmosphere. Perhaps the most significant bottleneck is the **lack of a well-developed in-space infrastructure.** The absence of reliable and cost-effective in-space transportation networks, refueling depots, and large-scale manufacturing capabilities means that we are still largely operating under a "one-way trip" paradigm, where everything needed in space must be launched from Earth.

Looking Ahead: Transitioning Beyond Earth Dependence

The limitations of the current Earth orbit supply chain clearly indicate the need for a fundamental shift towards a more self-sufficient model that leverages resources available in space. The following chapters will delve into the exciting possibilities of establishing supply chains based on resources found on the Moon, Mars, and within the asteroid belt. These extraterrestrial resources, such as water ice on the Moon and Mars (which can be used to produce propellant), regolith for construction, and the vast mineral wealth of asteroids, hold the key to breaking free from our dependence on Earth launch for all space activities. The development of in-space infrastructure, including orbital transfer vehicles, refueling depots strategically located in cislunar space, and in-space manufacturing facilities, will be crucial for enabling this transition. By exploring these alternative supply chain models, we can begin to envision a future where humanity's presence in space is not just a fleeting endeavor but a sustainable and thriving expansion into the cosmos.

PART II: Expanding the Supply Chain Beyond Earth

Like our ancestors we are lured to explore and expand our civilizations. From ancient sailors moving East across the Pacific Ocean, to traders trudging through the sands of the Middle East we have moved, traded and created new routes of travel. Every new route required initial scouting followed by settlers. Our next adventure is utilizing this same pattern. Only this time we have to bring all our supplies with us-even our air.

Figure 4.Author's rendition of the Earth to Moon travel.

Chapter Three: Lunar-Based Supply Chains: A Stepping Stone to the Solar System

The Moon as a Strategic Location

The introduction this book started with imagining waking up in a Lunar habitat. Breathing air cycled through filters both manufactured and natural (i.e. plants). Drinking water obtained from Lunar ice as well as the recycling center. Eating food produced from the same plants that clean the air but also from growth vats. The thick walls of your habitat are built from Lunar concrete using regolith in place of sand and rock. Your "window" is really a screen on the wall where you flip between the habitat news channel, streaming images of the Lunar surface, recorded scenes of Earth or even your favorite shows beamed from Earth orbit satellites.

Humanity's gaze has long been drawn to the Moon, our closest celestial neighbor. Beyond its poetic allure and historical significance as the first extraterrestrial body humans have walked upon, the Moon holds immense strategic value as the logical next step in establishing a robust space-based supply chain that extends beyond the confines of Earth orbit. Its unique combination of proximity, readily available resources, a significantly lower gravity well, and its potential as an ideal testbed for deep space technologies make the Moon an indispensable cornerstone in our multi-planetary future.

The **proximity** of the Moon to Earth is a paramount advantage. At an average distance of approximately 384,400 kilometers (roughly 238,900 miles), the travel time to the Moon is significantly shorter compared to other potential destinations like Mars or the asteroid belt. Crewed missions to the Moon typically take a few days, while cargo missions can be accomplished within a similar timeframe depending on the specific trajectory and propulsion systems used. This relative closeness has profound implications for supply chain management. Shorter transit times translate to reduced mission durations, lower exposure to the hazards of deep space, and quicker turnaround times for resupply missions. In the event of emergencies at a lunar base or during lunar surface operations, the proximity allows for relatively rapid return journeys to Earth, a critical factor for the safety and well-being of lunar inhabitants. Furthermore, the lower energy requirements (measured in terms of Delta-v, or change in velocity) to reach the Moon compared to more distant destinations make it an economically more feasible starting point for developing and testing the fundamental technologies and operational procedures required for a space-based supply chain. The Moon's proximity positions it as the natural proving ground for the intricate logistics and resource management systems that will eventually be necessary for more ambitious interplanetary endeavors.

Beyond its strategic location, the Moon boasts a wealth of **abundant resources** that can be harnessed to create a self-sustaining lunar supply chain and potentially support activities elsewhere in the solar system. Among the most critical of these resources is **water ice**. Extensive research, particularly through missions like NASA's Lunar Crater Observation and Sensing Satellite

(LCROSS), has confirmed the presence of significant quantities of water ice, primarily concentrated in the permanently shadowed regions (PSRs) at the lunar north and south poles (Anderson, 202; Choudhury, 2023; Kuthenur,2023). These regions, shielded from direct sunlight, maintain extremely low temperatures, allowing water ice to persist for billions of years. The importance of water ice in establishing a lunar supply chain cannot be overstated. It is essential for human life support, providing drinking water and serving as a feedstock for producing breathable oxygen through electrolysis. Crucially, water ice can also be split into its constituent elements, hydrogen and oxygen, which are highly efficient and powerful rocket propellants. The potential for a lunar propellant depot, fueled by locally sourced water ice, represents a paradigm shift in space transportation. Missions departing from such a depot would require significantly less propellant to be launched from Earth, dramatically reducing mission costs and increasing payload capacity for deep space exploration. Water ice, therefore, forms a fundamental "resource" element of a lunar supply chain, enabling both sustenance and transportation.

Another readily available lunar resource is **regolith,** the ubiquitous layer of loose, fragmented rock and dust that blankets the lunar surface. Regolith, formed by billions of years of meteorite impacts, is composed of various minerals and offers a multitude of potential uses within a lunar supply chain. Its most immediate application lies in its potential as a **construction material.** Lunar regolith can be sintered (heated to a high temperature without melting) to create bricks, tiles, and other structural elements that could be used for building habitats, radiation shielding, roads, and landing pads. Its density makes it an effective material for shielding against harmful solar and cosmic radiation, a significant concern for long-duration lunar missions. Furthermore, lunar regolith contains a significant amount of oxygen chemically bound within its minerals. Various technologies are being developed to extract this oxygen, which is vital for both life support and as the oxidizer component of rocket propellant. Regolith, therefore, represents the readily available "raw materials" for building infrastructure and producing a key life support element within a lunar supply chain.

While currently a subject of ongoing research and development, **Helium-3** is another lunar resource with potential long-term significance. This light, non-radioactive isotope of helium is relatively rare on Earth but is believed to be more abundant in the lunar regolith, deposited by the solar wind over eons. Helium-3 is considered a promising fuel for future fusion reactors (Berger, 2024; Sanshua, 2025), which could potentially provide a clean and virtually limitless energy source. The extraction of Helium-3 from lunar regolith and its potential transport back to Earth represent a long-term vision for a lunar-based supply chain contributing to terrestrial energy needs, although significant technological and economic hurdles remain.

Finally, lunar rocks and regolith contain various **metals** such as iron, aluminum, titanium, and silicon. While the concentration of these metals is generally lower than in terrestrial ores, their presence offers the potential for **in-situ manufacturing** of tools, equipment, and structural components on the Moon. Extracting and refining these metals locally could significantly reduce the reliance on launching heavy metallic components from Earth, further enhancing the self-sufficiency of a lunar supply chain, and creating the blueprints for a Martian supply chain. The development of efficient metal extraction techniques from lunar materials is a crucial step towards establishing a more robust lunar industrial base.

Beyond its resources, the Moon's **lower gravity well** presents a significant advantage for establishing a space-based supply chain. With a surface gravity approximately one-sixth that of Earth, the energy required to launch from the lunar surface is considerably less. This lower escape velocity makes it significantly easier and more cost-effective to transport lunar resources, such as water ice or extracted metals, to lunar orbit, Earth orbit, or even further into the solar system. The reduced energy expenditure translates directly into lower transportation costs within a lunar supply chain, making the Moon an attractive location for sourcing and distributing space-based resources.

The Moon also serves as an **ideal testbed for deep space technologies** and operational procedures that will be essential for more challenging destinations like Mars. Establishing a sustained human presence on the Moon and developing a functional lunar supply chain will provide invaluable experience in areas such as closed-loop life support systems, advanced robotics and automation, autonomous systems for resource extraction and processing, and long-duration human habitation in a harsh extraterrestrial environment. The lessons learned and technologies validated on the Moon will directly inform and accelerate our efforts to establish similar supply chains on Mars and other celestial bodies, making the Moon a critical stepping stone in our broader space exploration and development strategy.

Lunar Surface Supply Chain Needs

To establish a permanent and thriving presence on the Moon, a comprehensive supply chain will be required to support a range of activities, from basic human habitation to scientific research and resource utilization. The needs of a lunar surface base can be broadly categorized as follows:

Habitation and Base Support

Sustaining a long-term human presence on the Moon will necessitate a reliable supply of essential resources and infrastructure. **Food** will initially need to be transported from Earth, likely in pre-packaged forms. However, for a truly sustainable lunar base, the development of **in-situ food production** capabilities will be crucial. Research is underway into various methods, including hydroponics and aeroponics in controlled environments, and even the potential for utilizing lunar soil (regolith) for agriculture, though this presents significant challenges due to its composition and lack of organic matter. **Water**, as discussed earlier, will be primarily sourced from lunar water ice through extraction and processing. Ensuring a continuous supply of purified water for drinking and other life support functions is paramount. **Oxygen**, vital for breathing, will also be derived from lunar resources, primarily through the electrolysis of water ice or potentially through the extraction of oxygen from regolith. **Shelter materials** will be needed to construct habitats that protect astronauts from the harsh lunar environment, including extreme temperature variations, radiation, and micrometeoroid impacts. Initially, inflatable or prefabricated habitats launched from Earth may be used, but the long-term goal is to utilize lunar regolith for construction, either by sintering it into bricks or using it as a shielding layer over habitats. **Power generation and storage** are critical for all lunar surface operations. Solar power will likely be the primary source, but the long lunar nights (approximately 14 Earth days) necessitate robust energy storage solutions, such as advanced batteries or fuel cells. The potential for future nuclear power on the Moon could provide a more consistent and reliable energy source. **Life support consumables**, such as air filters, carbon dioxide scrubbers, and hygiene products, will need to be regularly supplied, with the eventual goal of developing closed-loop systems that recycle and regenerate these consumables in-situ. **Scientific equipment** will be essential for conducting research in various fields, including lunar geology, astrophysics, and the effects of long-duration effects of off world living conditions. The supply chain for this equipment will involve launching instruments from Earth and potentially manufacturing or repairing simpler components on the Moon. **Rovers**, both pressurized and unpressurized, will be necessary for transporting astronauts and equipment across the lunar surface, facilitating exploration and construction activities. The supply chain for rovers will include launching them from Earth and potentially developing the capability for in-situ maintenance and repair including locally manufactured replacement parts. Finally, **construction materials**, beyond those used for shelter, will be needed for building landing pads, roads, and other infrastructure to support a growing lunar base. Utilizing lunar regolith will be key to minimizing the reliance on Earth-launched construction materials.

ISRU (In-Situ Resource Utilization) on the Moon

The key to establishing a sustainable and cost-effective lunar supply chain lies in the successful implementation of In-Situ Resource Utilization (ISRU). **Water ice extraction and processing** will be a cornerstone of lunar ISRU. This reliance on moon ice will prove pivotal in geographic base locations Various methods are being proposed and tested to locate and extract water ice from the permanently shadowed regions. These include robotic excavators that can dig into the icy regolith and drills that can bore into deeper deposits. Once extracted, the icy regolith will need to be processed to separate the water ice and then purified. The purified water can then be electrolyzed to produce hydrogen and oxygen, the two components of highly efficient rocket propellant. The ability to produce propellant on the Moon would dramatically reduce the cost and complexity of lunar surface missions and enable more ambitious deep space endeavors. Storing cryogenic propellants like liquid hydrogen and oxygen in the harsh lunar environment presents its own set of challenges that will need to be addressed. **Regolith utilization** offers numerous opportunities for creating a more self-sufficient lunar base. As mentioned earlier, lunar regolith can be sintered using solar or microwave energy to create strong and durable construction materials like bricks and tiles. It can also be used as a radiation shield by layering it over habitats and storage facilities for liquid oxygen and hydrogen. Various chemical processes are being developed to extract oxygen from lunar regolith. Once these processes are matured they will allow for base expansion across the lunar surface. **Metal extraction** from lunar regolith is a more complex endeavor that is still in the early stages of research. The potential to extract metals like iron, aluminum, and titanium from lunar ores could eventually enable the in-situ manufacturing of tools, equipment, and even structural components, further reducing the reliance on Earth-based supplies.

Lunar Manufacturing and Production

Beyond simply extracting and utilizing raw resources, the establishment of **lunar manufacturing and production** capabilities will be crucial for creating a truly robust lunar supply chain. **3D printing,** also known as additive manufacturing, holds immense potential for producing tools, spare parts, and even structural components on demand using both imported and lunar-derived materials. This would significantly reduce the need to pre-ship a vast inventory of spare parts from Earth. The possibility of creating basic habitats or modules using lunar regolith as the primary raw materials for in-situ manufacturing techniques is also being explored. Establishing lunar manufacturing will require reliable and abundant sources of energy, as well as a high degree of automation to operate effectively in the lunar environment. Lunar manufacturing represents a significant step towards "closing the loop" in the lunar supply chain, enabling the creation of value and reducing dependence on Earth for manufactured goods.

Lunar Orbit Supply Chain Needs

In addition to the needs of a lunar surface base, a robust supply chain will also be required to support activities in lunar orbit. This will involve establishing infrastructure and transportation networks to facilitate movement between Earth, lunar orbit, and the lunar surface. It will be a transition from a raw lunar surface to a human civilization.

The **Lunar Gateway**, or similar orbiting stations, will serve as a critical hub in the lunar orbit supply chain. These stations are envisioned as multi-purpose platforms supporting transportation to and from the lunar surface, providing a communication relay between Earth and lunar operations, serving as a science platform for conducting research in lunar orbit, and potentially acting as a staging point for future missions to Mars and other deep space destinations. The supply chain needs for building, operating, and maintaining a lunar orbiting station like the Gateway will be significant. Initially, the various modules of the station will need to be launched from Earth. Once operational, regular resupply missions will be required to deliver consumables for crewed missions, transport scientific equipment and payloads, and potentially deliver propellant for refueling spacecraft. The ability to utilize lunar-derived propellant to refuel spacecraft at the Gateway would significantly enhance the efficiency and sustainability of the entire cislunar transportation network.

Efficient **cislunar transportation** systems will be essential for moving people and cargo between Earth orbit, lunar orbit, and the lunar surface. This will require the development of advanced **lunar**

landers capable of transporting crew and significant amounts of cargo between lunar orbit and the surface. These landers will need to be highly reliable and reusable to minimize costs. **Orbit Transfer Vehicles (OTVs)** will also be necessary for moving payloads and potentially crew between different orbits around Earth and the Moon. These vehicles could utilize various propulsion technologies, including chemical rockets for high-thrust maneuvers and electric propulsion for more efficient long-duration transfers. The supply chain for cislunar transportation will involve the manufacturing and assembly of these vehicles (initially likely on Earth, but potentially on the Moon in the future), the provision of propellant (with a strong emphasis on utilizing lunar-derived sources), and the development of in-space maintenance and servicing capabilities, potentially using robotic spacecraft.

A key element of a future lunar orbit supply chain is the concept of **fuel depots in cislunar space.** Strategically located fuel depots, perhaps in low lunar orbit or at stable Lagrange points, could be supplied with propellant derived from lunar water ice. These depots would then serve as refueling stations for spacecraft traveling to the Moon, Mars, and beyond. The benefits of such a system are numerous. Spacecraft departing from Earth orbit would not need to carry the full amount of propellant required for their entire mission, significantly increasing their payload capacity. Missions to the lunar surface could be staged from a fuel depot in lunar orbit, reducing the amount of propellant needed for the initial Earth launch. Furthermore, fuel depots would enable more complex and ambitious deep space missions by providing refueling opportunities along the way. Establishing and operating fuel depots will require developing the infrastructure to store large quantities of cryogenic propellants in the vacuum of space for extended periods, as well as efficient and safe propellant transfer technologies. The ability to create and utilize fuel depots in cislunar space represents a critical step towards a more sustainable and affordable space transportation supply chain. A potential long term technological leap would allow for a lunar space elevator. This elevator would connect an orbital platform directly to a transit point (or possibly multiple points) on the lunar surface. The concept is well understood but the technology to build the tether strong enough to withstand the tension, environmental impacts and allow for elevator movement does not exist. The projected reduction in transport costs and transport time to and from orbit will make this investment critical but likely not pursuable before significant lunar infra structure development.

Figure 5. Lunar base space elevator concept. Retrieved from Wikipedia commons on April 6th 2025.

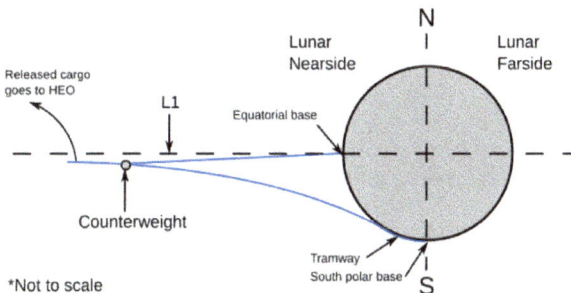

Supporting Earth Orbit Needs from the Moon

In the long term, a mature lunar-based supply chain could potentially extend its reach to support activities in Earth orbit, further reducing our reliance on costly and environmentally impactful Earth launches. The progression from existing technology to process regolith and moon ice forward to extractions of minerals, oxygen and helium 3 from the regolith will act as stepping stones. Each step furthers humanities ability to shift from heavy reliance and Earth based supply chains, to ISRU self-reliance and finally to a Moon based supply chain supporting off moon demand.

One of the most significant ways the Moon could support Earth orbit is by supplying **lunar propellant for Earth orbit missions.** As discussed earlier, the Moon has the potential to become a significant source of rocket propellant in the form of liquid hydrogen and oxygen derived from water ice. Transporting this lunar-derived propellant to Earth orbit could offer significant economic advantages for certain types of missions. For example, upper stages of rockets used to send satellites to geostationary orbit could be refueled in low Earth orbit with lunar propellant, reducing the overall size and cost of the launch vehicle needed from Earth. Similarly, satellites in high Earth orbits that require periodic orbital correction boosts could be refueled in orbit using lunar propellant, extending their operational lifespan. Even deep space missions originating from Earth orbit could potentially benefit from refueling with lunar propellant in Earth orbit, increasing their payload capacity for exploration. Establishing a supply chain to transport lunar propellant to Earth orbit would involve a complex series of steps, including extraction and processing on the Moon, transportation from the lunar surface to cislunar space, transfer to Earth orbit using specialized OTVs, and the development of infrastructure for storing and distributing propellant in Earth orbit. A central flight planning and oversight system needs to be developed to reduce crash risks between a perceived increase in space vehicle traffic. While the initial investment in such a system would be substantial, the long-term cost savings compared to launching all propellant from Earth could be significant.

In the more distant future, there are also possibilities for supplying Earth orbit with **lunar-derived materials.** While the high cost of transporting bulk materials from the Moon to Earth orbit currently makes this economically unviable for most applications, there might be niche areas where specialized materials or components produced on the Moon could be valuable. For example, certain metals processed in the low-gravity environment of the Moon might have unique properties that make them desirable for specific applications in Earth orbit. Similarly, radiation-shielding materials produced from lunar regolith could potentially be more cost-effective to transport to Earth orbit than launching equivalent materials from Earth. However, realizing this potential will require significant advancements in transportation technology and a substantial reduction in the cost of moving mass between the Moon and Earth orbit.

Challenges and Opportunities for Lunar Supply Chains

The development of a robust lunar-based supply chain presents a multitude of exciting opportunities, but it also faces significant technological and logistical challenges. **Technological hurdles** abound. We need to further develop and test reliable and efficient technologies for **ISRU** in the harsh lunar environment, including methods for extracting water ice, processing regolith, and potentially extracting metals. **Lunar surface operations** will need to become more efficient and resilient, addressing challenges such as dust mitigation, dealing with extreme temperature variations, and operating during the long lunar nights. Building a robust **transportation infrastructure** for cislunar space and the lunar surface, including reusable landers and orbit transfer vehicles, is also critical. Finally, the long-term storage and transfer of **cryogenic propellants** in the vacuum and temperature extremes of space present significant engineering challenges that need to be overcome.

The **economic viability** of a lunar supply chain will be a key factor in its success. We need to carefully analyze the costs and benefits of utilizing lunar resources compared to relying solely on Earth-based supplies. The demand for lunar resources and services, as well as the potential for generating revenue from these activities, will play a crucial role in attracting public and private investment. The development of new markets and economic opportunities related to lunar activities will be essential for the long-term sustainability of a lunar supply chain. Initial investments will likely be driven by Governments transitioning to more commercial activities and the infrastructure matures.

International cooperation will likely be a significant aspect of establishing lunar supply chains. Sharing resources, developing common infrastructure, and collaborating on research and development efforts could help to reduce costs and accelerate progress. However, geopolitical considerations and the need for international agreements on the utilization of lunar resources will also need to be addressed. The current legal environment leaves a lot more unknowns than knowns which presents both risks and opportunities.

Despite these challenges, the **immense potential** of a lunar-based supply chain is undeniable. It offers a pathway to more affordable and sustainable space exploration, enabling a permanent human presence on the Moon and providing the resources needed for more ambitious missions to Mars and beyond. The development of lunar resources could also potentially create new economic opportunities both in space and back on Earth. The Moon stands as a crucial stepping stone in humanity's journey to becoming a multi-planetary species, and establishing a robust lunar supply chain is a fundamental requirement for realizing this ambitious vision.

Chapter Four: Martian Supply Chains: Enabling Human Presence and Resource Return

Figure 6: Author's rendition of travel from Earth to Mars.

Mars: A More Complex Supply Chain Challenge

Let's take the next step on this journey to new worlds. Imagine going to work at the Martian habitat power station. Solar power collection on the surface is challenging due to the rapid build up of dust. The more effective alternative is the micro nuclear reactor set up just out side the main facility. It provides all the electricity needed to power the lights, heat, air flow, food processing, launch sites, manufacturing, medical station, science labs, mining machines, entertainment and more. The habitat couldn't exist without it and you enjoy knowing you help keep all this smoothly operating. Today though is not a fun day. You are assigned to inspect the power cable housings from the outside. A day of trudging through Martian dust in your space suit to double check what the drones might have missed.

While the Moon presents a significant step beyond Earth orbit in terms of establishing a space-based supply chain, Mars represents a far more complex and ambitious undertaking. Similar to long distance sailing voyages, travelers to Mars must bring tremendous amounts of supplies. The trips are also not direct lines. Just as a sailing ship must account for curvature of the Earth, travelers to Mars must plan to meet their destination at a future position in orbit. The challenges associated with creating a sustainable and functional supply chain on the Red Planet are magnified by the **greater distance** from Earth, a significantly **harsher environment**, and the necessity for **longer mission durations**. Understanding these fundamental differences is crucial for appreciating the scale of the logistical and technological hurdles that must be overcome to enable a permanent human presence and potentially even the return of Martian resources to Earth. A trip to Mars is very closely aligned to submarines headed to the bottom of the Ocean.

The **greater distance** to Mars, which varies significantly depending on the relative positions of the two planets in their orbits, presents a substantial increase in transit times compared to the Moon. At its closest approach, Mars is still tens of millions of kilometers away, resulting in travel times for crewed missions that can range from several months to nearly a year, depending on the propulsion systems and trajectory used. Cargo missions, often utilizing more efficient but lower-thrust propulsion like solar electric propulsion, can take even longer. This extended transit time has profound implications for supply chain management. Resupply missions will be far less frequent than to the Moon, requiring a much greater degree of self-sufficiency on the Martian surface. Emergency return scenarios become significantly more challenging and time-sensitive. The sheer duration of interplanetary voyages to Mars also necessitates careful consideration of crew health, both physical and psychological, and the need for robust onboard medical capabilities as part of the overall life support system. Items such as saline IVs, or blood transfers are extremely challenging to resupply within small populations far from Earth. From a supply chain perspective, the longer distances translate to increased logistical complexity, greater reliance on pre-positioned supplies, and the need for highly reliable systems with minimal maintenance requirements. The "just-in-time" delivery models that might be conceivable for lunar operations become virtually impossible for Mars, demanding a more robust and redundant approach to resource management.

The **harsh** environment of Mars further complicates the establishment of a sustainable supply chain. Unlike the Moon, Mars possesses a thin atmosphere, primarily composed of carbon dioxide (Hu et al., 2013; Murray & Jagoutz, 2024), with a surface pressure less than 1% of Earth's. This thin atmosphere offers minimal protection from solar and cosmic radiation and contributes to extreme **temperature variations** between day and night, often exceeding 100 degrees Celsius. The lack of a global magnetic field, coupled with the thin atmosphere, means that the Martian surface is bombarded by significantly higher levels of **radiation** than Earth, posing a serious health risk to unprotected humans and potentially damaging sensitive equipment. Unlike the Moon Mars is prone to massive **dust storms** that can engulf the entire planet for weeks or even months. These storms can severely impact solar power generation, reduce visibility to near zero, and potentially damage or impede the operation of surface equipment. The Martian **soil composition** also presents challenges. It contains perchlorates (Rymski, et al., 2024), which are toxic to humans and can complicate efforts to grow food in Martian regolith without proper treatment. The presence of these perchlorates (Rymski, et al., 2024) also needs to be considered in the context of ISRU efforts. This significantly more challenging environment necessitates a Martian supply chain that can provide much more robust and complex life support systems, more effective radiation shielding, and equipment designed to withstand extreme temperatures and dust ingress.

The necessity for **longer mission durations** on the Martian surface further underscores the complexity of the supply chain challenge. Due to the orbital mechanics of Earth and Mars, optimal launch windows for interplanetary travel occur roughly every 26 months. This synodic period dictates that crewed missions to Mars will likely involve extended stays on the Martian surface, potentially lasting for hundreds of days to align with the next favorable launch window for a return journey. Such long durations place immense demands on the Martian supply chain. **Closed-loop life support systems** capable of recycling air, water, and waste with extremely high efficiency will be essential to minimize the reliance on Earth resupply. **In-situ resource utilization (ISRU)** will be absolutely critical for producing key consumables like water, oxygen, and propellant on Mars to sustain a long-term human presence. The longer mission durations also necessitate larger habitats to accommodate crews for extended periods and more sophisticated medical capabilities to address potential health issues that may arise far from Earth's medical infrastructure. From a supply chain perspective, the need for extended surface operations amplifies the importance of reliability, redundancy, and the ability to repair and maintain equipment using Martian resources.

Mars Supply Chain Needs

Sustaining a permanent human presence on the Martian will require a highly sophisticated and largely self-sufficient supply chain. The needs of a Martian base will be significantly more complex than those of a lunar outpost due to the more challenging environment and longer mission durations.

Figure 7: Author's rendition of a subsurface Martian base.

Sustaining a Martian Base/Habitat will demand **a system more akin to a micro eco system** than those envisioned for the Moon. Maintaining a breathable atmosphere within pressurized habitats will require advanced systems for regulating pressure, temperature, and the composition of the air. Highly efficient **water recycling** and purification systems will be crucial to conserve this precious resource. Robust **waste management and recycling** processes will be necessary to minimize the volume of waste that needs to be stored or processed. **Food production** on Mars will likely involve a combination of initially transported food and the development of sophisticated **in-situ agriculture**. This could involve closed-environment hydroponic or aeroponic systems, and potentially even the cultivation of crops in treated Martian regolith under artificial lighting. The development of **closed-loop bioregenerative life support systems**, which integrate biological components like algae to produce food and oxygen while recycling waste, represents a long-term goal for Martian sustainability. Effective **radiation shielding** will be paramount to protect astronauts from the harmful radiation environment. This could involve burying habitats beneath layers of Martian regolith or using specialized radiation-resistant materials. Reliable **power generation** will be essential to support all base operations. While solar power will likely play a role, the potential for planet-wide dust storms suggests that other power sources, such as nuclear power, may be necessary for a consistent and dependable energy supply. **Larger habitats** will be required to accommodate crews for extended durations, providing adequate living and working space to mitigate the psychological challenges of isolation. These habitats may involve modular designs or inflatable structures. **Advanced rovers**, both pressurized for long-distance traverses and unpressurized for shorter excursions, will be vital for exploration, construction, and transporting resources. The use of autonomous robotic swarms for various tasks will also be critical. Since early missions are most likely going to be Government funded, a comprehensive supply of **scientific equipment** will be needed to conduct research on Mars, and robust **construction equipment** will be required to build and maintain the base infrastructure.

Advanced ISRU on Mars will be absolutely critical for establishing a sustainable Martian supply chain. The Martian atmosphere, being primarily composed of carbon dioxide (Hu et al., 2013; Murray & Jagoutz, 2024), offers a valuable resource for propellant production through the **SABATIER/Methane Reaction**. This process involves reacting atmospheric CO_2 with hydrogen (which can be derived from water ice) to produce methane and water. Methane and oxygen, also derivable from water or atmospheric CO_2, form a highly efficient propellant combination for the Martian environment, offering a higher specific impulse in Mars' thin atmosphere (Schultz, 2014) compared to hydrogen and oxygen. Extracting and utilizing **water ice** will be another cornerstone of Martian ISRU. Evidence suggests the presence of significant subsurface water ice deposits (Watters et al., 2024), particularly at higher latitudes. Developing reliable methods for locating and extracting this ice, such as drilling and melting, will be essential for providing drinking water, life support consumables, and raw material for propellant production through electrolysis. Martian water could also potentially be used for **hydroponics and other forms of agriculture**, further enhancing the base's self-sufficiency. Utilizing **Martian regolith** for **construction materials** will be crucial for building habitats and infrastructure without relying on massive shipments from Earth. Techniques such as sintering regolith into bricks or using it as a raw material for additive manufacturing (3D printing) could provide the necessary building blocks. Research is also underway to explore the potential for creating cementitious materials from Martian resources. Extracting **oxygen** directly from the Martian atmosphere through processes like direct CO_2 capture and electrolysis could provide another vital resource for life support and propellant.

Manufacturing and Fabrication on Mars will play a vital role in creating a more self-sufficient Martian supply chain. **Advanced 3D printing** technologies will be essential for producing tools, spare parts, and even components for habitats and equipment on demand, using both materials brought from Earth and those derived from Martian resources. **Automated construction** using robots and Martian regolith will be necessary to build and expand the base infrastructure over time. The initial stages of a potential **Martian industry** could focus on producing basic consumables or materials needed for the base, such as simple polymers or processed regolith for construction.

Mars Orbit (and Phobos/Deimos) Supply Chain Needs

Establishing a robust Martian supply chain will also necessitate infrastructure and capabilities in orbit around Mars and potentially on its moons, Phobos and Deimos. **Orbital infrastructure around Mars** will be crucial for various functions. **Communication relay satellites** will be needed to ensure reliable communication between Earth, Mars orbit, and the Martian surface, overcoming the challenges posed by the planet's rotation and terrain. **Observation platforms** in Martian orbit will provide valuable data for continuous monitoring of the planet's weather patterns, climate, and surface changes, as well as for reconnaissance of potential landing sites and resource locations for future missions. The possibility of establishing **orbital assembly and staging areas** could also be explored, where large spacecraft for further exploration of the Martian system or for voyages to other destinations could be assembled and prepared.

The Martian moons, **Phobos and Deimos**, offer intriguing possibilities as potential resource outposts within the overall Martian supply chain. Their **proximity to Mars** and their **low gravity** make them potentially attractive locations for certain activities. There is evidence to suggest the presence of **water ice** on these moons (Cordell, 1985; Fanale & Savail, 1989), particularly on Deimos based on spectral analysis. Their low escape velocity could make them ideal locations for establishing **propellant depots** that could be supplied from the Martian surface or potentially from resources found on the moons themselves. Phobos and Deimos could also serve as **staging points** for missions to other locations within the Martian system or even as initial **resource processing sites** where raw materials could be partially processed before being transported to the Martian surface or orbit, taking advantage of their weaker gravitational pull. A Martian space elevator would require a significant leap in technology and SITU manufacturing. The benefits would be equally profound to a Moon based system but far more challenging in terms of gravitational effects and tether length.

Efficient **interplanetary transportation** to and from Mars will be a fundamental requirement for the entire Martian supply chain. Developing more efficient interplanetary transfer vehicles is crucial for reducing transit times and increasing payload capacity. **Nuclear Thermal Propulsion (NTP)** offers the potential for significantly faster transit times compared to traditional chemical rockets. Early missions could convert a landed space craft's NTP system into a local power source. **Advanced chemical rockets** with improved specific impulse and thrust could also contribute to shorter travel durations. **Solar Electric Propulsion (SEP),** while offering lower thrust and longer transit times, could be highly efficient for transporting large cargo payloads. Other more advanced propulsion concepts, such as fusion propulsion or beamed energy, may become relevant in the longer term. The development of **reusable interplanetary vehicles** will be essential for reducing the overall cost of repeated missions to Mars. The transportation of both **crewed and cargo missions** will require different considerations and potentially different vehicle designs, tailored to the specific needs of each type of mission. The potential thrust capacity of a fusion drive would lead to potential speeds that could shorten the trip from months to days. Unfortunately, like the space elevators this concept is still far ahead of capability. Similar to the space elevator this type of demand can be a driver of both private and public investment.

Return Transport of Precious Metals to Earth (Long-Term Vision)

While the primary focus of establishing a Martian supply chain will initially be on supporting human presence and exploration, the long-term vision could potentially include the **return transport of precious metals to Earth.** Identifying **Martian resources of economic value** that could justify the immense cost and effort of returning them to Earth is a key aspect of this long-term vision. Scientific evidence suggests the potential presence of **rare earth elements (REEs)** and **platinum group metals (PGMs)** on Mars, based on geological data collected by rovers and the analysis of Martian meteorites found on Earth. These elements are highly valuable for various industrial and technological applications on Earth. However, our current understanding of the concentration and accessibility of these resources on Mars is still limited and largely speculative. Further exploration and resource assessment will be needed to determine if economically viable deposits exist, and if so, are they only economically viable in-situ.

The **supply chain for resource extraction, processing, and return** from Mars to Earth would be incredibly complex and challenging. It would involve sophisticated **mining operations** on the Martian surface, potentially utilizing highly automated equipment to cope with the harsh environment. Advanced **refining processes** would be required to extract and concentrate the valuable metals from the Martian ores, potentially requiring significant energy input and specialized chemicals that might need to be manufactured in-situ or transported from Earth. Once processed, the resources would need to be **launched from the Martian surface** into orbit, requiring dedicated launch vehicles adapted to the Martian environment. The journey back to Earth would involve **interplanetary transport**, with transit times of several months. Finally, upon reaching Earth, the returning spacecraft would need to safely undergo **entry into Earth's atmosphere** and land, requiring robust heat shields and landing systems. The entire process, from initial resource extraction on Mars to the arrival of the refined materials on Earth, would be a long-term and incredibly resource-intensive undertaking, requiring significant technological advancements and a compelling economic justification.

Challenges and Innovations for Martian Supply Chains

Establishing a functional and sustainable supply chain on Mars presents a multitude of **extreme challenges**. The harsh **radiation** environment necessitates robust shielding for both habitats and astronauts. Frequent and potentially planet-wide **dust storms** can severely disrupt operations and solar power generation. The extreme **temperature variations** require equipment and habitats designed to withstand a wide range of thermal conditions. The **long transit times** to and from Mars pose significant logistical and human health challenges. For all human activities, the **psychological factors** associated with long-duration isolation and confinement in a remote and alien environment must be carefully considered.

Overcoming these challenges will require significant **technological breakthroughs. Advanced ISRU** technologies capable of reliably and efficiently producing propellant, water, and other essential resources from Martian materials are paramount. **Closed-loop life support systems** with extremely high recycling rates will be crucial for long-duration missions. The development of highly **autonomous systems** and robotics will be essential for exploration, construction, maintenance, and resource processing with minimal human intervention. **Robust habitats** capable of withstanding the Martian environment and providing adequate radiation shielding are needed. Finally, the development of more efficient and faster **advanced propulsion systems** will be critical for reducing transit times and increasing payload capacity for interplanetary travel.

The establishment of a Martian supply chain will require a **long-term commitment** of significant financial resources and sustained effort from both governments and the private sector. **International collaboration** will likely be essential to share the costs and risks associated with this ambitious undertaking. A clear and long-term **strategic vision** will be needed to guide the development of the necessary technologies and infrastructure. Although not a direct component of supply chains, we must also address the **ethical considerations**, including the potential for planetary protection and the broader implications of human expansion into the cosmos. Multiple colonies could exist independent of each other on the Mars surface for many years before potential conflicting interests arise. Sooner or later though, clear understanding of legal structure will be pertinent to maintaining the systems and not disrupting supply chains.

Chapter Five: Asteroid Belt Supply Chains: Resource Abundance and Deep Space Logistics

The Asteroid Belt: A Vast Reservoir of Resources

Now we venture to our final destination, the belt. Here survival is still a daily activity. Technology makes it possible but human habitats this far out are small and austere. You wake up tethered to the wall so you don't float away in the zero gravity. Breakfast is a prepacked meal in a tube, as is the water which you need to ration. You float around the spacecraft conducting your morning routine of checks, first the air filters, then the power drive, then your food and water stores and finally the scopes for and communication systems. Failures in any of these areas can prove fatal to you, but soon the new space port construction will be complete and you will have a real habitat to stay in. At least on temporary trips.

The asteroid belt, a torus-shaped region in the Solar System located roughly between the orbits of Mars and Jupiter, represents a treasure trove of raw materials on a scale almost unimaginable. Containing millions, and potentially trillions, of individual asteroids ranging in size from pebbles to dwarf planets like Ceres, the total estimated mass of this celestial body exceeds that of our own Moon. This makes the asteroid belt not just a collection of space rocks, but a vast, untapped resource depot capable of fueling a burgeoning space economy and supporting human expansion throughout the solar system. It is the heart of a new gold rush.

The **immense potential** of the asteroid belt lies in its sheer scale. Unlike the finite and often politically contested resources on Earth, or even the more limited accessible resources on the Moon and Mars, the asteroid belt offers a seemingly inexhaustible supply of materials. This vastness positions it as a critical long-term solution for the resource needs of a growing space-faring civilization, capable of sustaining large-scale construction projects, powering deep space missions, and potentially even alleviating resource scarcity on Earth. It is realistic to imagine that long before any replacement technologies alleviate the need for space mining, we will not have come close to exhausting the abundance of the asteroid belt.

Figure 8. Author's visual depiction of asteroid mining.

The **variety of resources** available within the asteroid belt is equally compelling. Among the most significant are:

Metals:

- **Iron and Nickel:** Abundant in M-type (metallic) asteroids (Margot & Brown, 2003), these are fundamental building blocks for space construction and manufacturing. Imagine habitats, spacecraft hulls, and large-scale infrastructure built with readily available iron and nickel sourced directly from the asteroid belt, drastically reducing the need for costly and energy-intensive launches from Earth. These metals can be transported to lunar orbit for habitat construction, to Martian orbit for building interplanetary transfer vehicles, or even processed in the asteroid belt to create modular components for future space based supply stations.

- **Cobalt:** A crucial component in high-performance batteries and specialized alloys, cobalt extracted from asteroids could support the energy storage systems for lunar bases, Martian rovers, and spacecraft operating throughout the solar system, creating a reliable energy supply chain independent of Earth.

- **Platinum Group Metals (PGMs):** Extremely rare and valuable on Earth, PGMs like platinum, palladium, and rhodium are found in significant concentrations in some asteroids. These metals are essential for electronics, catalytic converters, and various high-tech applications. Returning even a fraction of these resources to Earth could revolutionize terrestrial industries, while in space, they could be used in advanced electronics for spacecraft and supply station systems, creating a high-value material flow within the solar system supply chain.

Water Ice:

- Primarily found in C-type (carbonaceous) asteroids (Tillman, 2017), water ice is a lifeblood resource for space exploration. It can be used directly for human consumption and life support on lunar and Martian bases, as well as in orbiting habitats and space based supply stations. More importantly, it can be electrolyzed into hydrogen and oxygen, the most efficient and widely used chemical rocket propellant. Asteroid-derived water ice can fuel lunar landers, Martian ascent vehicles, and interplanetary transfer stages, creating in-space refueling depots that dramatically reduce the cost and complexity of deep space missions originating from Earth or other locations.

Volatile Compounds:

- C-type asteroids (Tillman, 2017) also contain significant amounts of ammonia, methane, and other volatile compounds. Ammonia can be used as a precursor for fertilizers in potential Martian or lunar agriculture, supporting food production within those supply chains. Methane, along with oxygen, forms another efficient rocket propellant combination, particularly well-suited for use in the outer solar system. These volatiles can also be used in various industrial processes in space, potentially for creating polymers or other useful materials, further diversifying the outputs of the asteroid belt supply chain.

The **different asteroid types and their resource distribution** have significant implications for prospecting and mission planning within the asteroid belt supply chain. **C-type asteroids** (Tillman, 2017), the most common type, rich in water ice, carbon, and organic compounds, are predominantly located in the outer regions of the main belt. Missions targeting water ice and volatiles will need to venture further from the inner solar system. S-type asteroids, the second most common, composed mainly of silicate rocks and some metals, are found primarily in the inner regions. These could be targeted for their silicate content for construction or for the metals interspersed within them. M-type asteroids (Margot & Brown, 2003), while relatively rare, contain the highest concentrations of valuable iron, nickel, and PGMs and are located in the middle regions of the main belt. Understanding this distribution is crucial for planning efficient prospecting missions and for strategically locating mining operations and future supply stations to minimize transportation distances and maximize resource acquisition.

Compared to **terrestrial resources**, the potential abundance of certain materials in the asteroid belt is staggering. For example, the amount of PGMs estimated to reside in asteroids far exceeds all known reserves on Earth. Similarly, the readily accessible water ice in C-type asteroids could dwarf the easily obtainable water resources (NASA, 2005; NASA 2025) on the Moon and potentially even Mars. This abundance offers the potential to alleviate resource scarcity on Earth for certain critical materials and, more importantly, to enable a truly self-sustaining and expansive space-based economy, where the building blocks for future infrastructure and exploration are sourced directly from space itself, rather than being constantly lifted from Earth's gravity well.

Asteroid Belt Supply Chain Needs

Establishing a functional and productive supply chain within the asteroid belt will require a sophisticated and interconnected network of robotic operations, in-space processing facilities, and long-distance transportation systems. **Resource prospecting and asteroid characterization** form the critical first step in this endeavor. **Developing robotic probes** capable of navigating the vast distances of the asteroid belt and operating autonomously for extended periods is essential. These probes will need to be equipped with advanced navigation systems to avoid collisions with the numerous asteroids and to accurately reach their target destinations. **Mission architectures** will vary depending on the objectives, ranging from flyby missions for initial reconnaissance to orbiting missions for detailed analysis of specific asteroids and even landing missions for in-situ sampling.

Sensor technologies will play a crucial role in identifying and analyzing the composition of asteroids from afar. Spectrometers can analyze the light reflected from an asteroid to determine its mineral composition. Radar can be used to map the surface and potentially detect subsurface water ice. Imaging systems will provide high-resolution views of asteroid surfaces, helping to identify potential mining sites. **Data analysis and target selection** will involve sophisticated algorithms and potentially AI to process the vast amounts of data returned by these probes and identify the most promising resource-rich asteroids for future exploitation.

The potential for deploying **Small Sat constellations** – large numbers of small, low-cost probes – could significantly enhance the speed and comprehensiveness of asteroid belt prospecting, providing a more detailed map of available resources. Connecting this crucial stage to the overall supply chain, prospecting and characterization represent the "identification and assessment" phase, providing the necessary information to guide subsequent mining and processing operations and inform the strategic placement of supply stations.

Once promising asteroids have been identified, **autonomous mining and extraction operations** will be necessary to access their valuable resources. This will involve envisioning fleets of robotic miners – specialized spacecraft capable of approaching, securely anchoring to, and extracting materials from asteroids. The design of these robotic miners will vary depending on the type and size of the asteroid and the resources being targeted. **Mining techniques** could include surface extraction for loose regolith or near-surface deposits, subsurface drilling for accessing deeper veins of metal or water ice, and even the more ambitious concept of capturing and processing entire small asteroids within a mobile processing facility. **Processing and sorting** will need to occur in-situ, at least to some extent, to separate the desired resources from the bulk material and prepare them for transport. This could involve crushing and grinding rock, melting ice, or using magnetic fields to separate metallic components. **Autonomous navigation and coordination** will be critical for operating large fleets of robotic miners across the vast distances of the asteroid belt. Sophisticated AI algorithms and reliable communication systems will be required to ensure efficient operation, collision avoidance, and coordinated efforts between multiple mining units. Considerations for **minimizing environmental impact** will also be important, even in the relatively pristine environment of the asteroid belt. Developing responsible mining practices that minimize disruption to asteroid dynamics and prevent the spread of contaminants will be crucial for the long-term sustainability of this resource. Connecting this stage to the supply chain, autonomous mining and extraction represent the "acquisition" phase, where raw materials are harvested from their celestial sources.

The raw materials extracted from asteroids will often require further processing and refinement in space to be truly useful. **In-Space Processing and Refining in the Asteroid Belt** will involve **developing efficient in-space techniques** for separating and refining valuable materials in the unique conditions of microgravity and vacuum. **Gravity-based separation challenges** will need to be overcome, as traditional methods relying on density differences are ineffective in the absence of significant gravity. **Alternative separation methods** will need to be employed, such as magnetic separation for metallic ores, chemical leaching using solvents to dissolve target materials, and electrolysis for separating elements from compounds. The **energy requirements** for these processes will be substantial, highlighting the need for efficient and reliable power sources, such as large solar power arrays or potentially future fusion reactors deployed within the asteroid belt. The development of **modular and scalable processing plants** will be essential, allowing for processing capabilities to be deployed to different asteroids as needed and scaled up as production increases. These processing plants could be located on dedicated spacecraft or potentially on larger, more stable asteroids or even purpose-built supply stations. Connecting this stage to the supply chain, in-space processing and refining represent the "transformation" phase, where raw, extracted materials are converted into usable forms for various applications.

Long-Distance Deep Space Transportation will be necessary to move resources within the asteroid belt and between the belt and other locations in the solar system, including Mars, the Moon, and Earth. This will require the development of highly efficient propulsion technologies. **Solar Electric Propulsion (SEP)** is a promising option for transporting large cargo masses over long distances, although transit times can be longer. **Advanced Chemical Rockets** may be necessary for faster transit times, particularly for crewed missions or time-sensitive cargo. Nuclear Thermal Propulsion (NTP) offers the potential for significantly faster and more efficient deep space travel compared to chemical propulsion. Ion propulsion is another efficient electric propulsion option suitable for long-duration missions. In the more distant future, **potential future concepts** like fusion propulsion or beamed energy could offer even greater efficiency and speed. **Autonomous navigation and trajectory planning** will be crucial for these long-distance voyages, requiring sophisticated systems to navigate the complex gravitational environment of the solar system and plan efficient trajectories over vast distances. The logistics of **cargo transfer and handling** in space will also need to be addressed, developing efficient methods for transferring resources between mining operations, processing facilities, supply stations, and transport vehicles. Connecting this stage to the supply chain, long-distance deep space transportation represents the "distribution" phase, moving the processed resources to where they are needed throughout the solar system.

To effectively support these complex operations and facilitate the flow of resources, **establishing permanent supply stations** in or near the asteroid belt will be crucial. These outposts will serve as critical logistical hubs, enabling a more efficient and sustainable asteroid belt supply chain. Permanent supply stations will be essential for providing a stable infrastructure to support the continuous operations of resource prospecting, mining, processing, and transportation within the asteroid belt. They will act as central nodes for logistics, refueling, maintenance, and potentially even crew support if human oversight is deemed necessary for certain complex tasks. Without such infrastructure, operations would be far more dispersed and inefficient, increasing transit times and overall costs.

Potential Locations

Several types of celestial bodies or man-made structures could serve as locations for these supply stations. **Nearby Moons are a viable option.** While the asteroid belt itself lacks large natural satellites, strategically located supply stations could potentially be established on moons of Jupiter or Saturn if those regions of the belt are found to be particularly resource-rich and accessible. These moons offer the advantage of a stable gravitational environment and could potentially provide additional resources themselves.

Large Asteroids such as the Dwarf planet Ceres, located in the inner asteroid belt, or other large asteroids like Vesta could serve as stable anchor points for constructing supply stations. Their larger size provides more surface area and potentially internal volume for habitats, processing plants, and storage facilities. Their gravity, while still low, could also aid in certain processing techniques.

Man-Made Space Stations present a combination of technology challenges and tailored crafting benefits. Purpose-built orbital facilities, constructed using asteroid-derived metals, could be positioned at strategic locations within the asteroid belt or at key orbital intersections. These stations could be designed with specific functionalities in mind, such as dedicated refueling depots, processing hubs, manufacturing, maintenance and repair facilities for robotic fleets.

Functions of Supply Stations

These permanent outposts would perform a variety of critical functions within the asteroid belt supply chain. They would serve as central logistics and storage locations for raw, partially processed, and fully refined resources extracted from various asteroids. This would allow for the accumulation of sufficient quantities of materials for efficient transport to other destinations. They can also serve as **refueling Depots supporting** transport vehicles operating within the asteroid belt and those making journeys to Mars, the Moon, or Earth. Propellant derived from asteroid water ice could be stored and transferred at these depots, significantly extending the range and capabilities of deep space missions. Given the vast distances and long operational times involved, supply stations would house advanced robotic manufacturing and repair facilities capable of servicing and maintaining the fleets of robotic miners and transport vehicles operating throughout the asteroid belt. This would minimize downtime and ensure the long-term functionality of the supply chain. While the initial stages of asteroid mining will likely be heavily reliant on autonomous systems, permanent supply stations could include habitats for human crews responsible for oversight, complex repairs, research, and the development of new technologies in-situ. These stations would also act as vital communication relays, facilitating reliable data transfer and command signals between Earth and the numerous robotic assets operating throughout the asteroid belt, overcoming the challenges of long communication delays.

Establishing and maintaining these permanent supply stations will itself require a dedicated supply chain. Initially, components and construction materials might need to be transported from Earth or the Moon. However, the long-term goal would be to utilize asteroid-derived metals and other resources to expand and maintain these stations, creating a self-sustaining infrastructure within the asteroid belt. This would involve transporting processed metals to the station construction sites and utilizing in-space assembly techniques, potentially with robotic assistance. Regular resupply missions from Earth or other locations might still be necessary for certain specialized components or consumables, but the reliance on external sources would gradually decrease as the asteroid belt infrastructure matures.

Potential Supply Chain Outputs from the Asteroid Belt

The establishment of a robust asteroid belt supply chain promises a wealth of potential outputs that can revolutionize space exploration and potentially impact terrestrial economies.

Metals for Space Construction and Manufacturing represent a primary output. The vast reserves of iron, nickel, and aluminum in M-type asteroids (Margot & Brown, 2003) offer the potential to **supply materials for building large, durable space habitats** in various locations throughout the solar system. Imagine constructing expansive orbital habitats in Earth orbit to alleviate population pressures or creating permanent settlements in lunar orbit or at stable Lagrange points to support further lunar development. These asteroid-derived metals could also be used for infrastructure development, such as building massive solar power arrays to beam energy to Earth or to power space-based industries, and constructing large communication platforms to improve connectivity throughout the solar system. Furthermore, the availability of these metals in space could revolutionize **spacecraft manufacturing**, allowing for the construction of larger and more complex spacecraft without the limitations and costs associated with launching massive structures from Earth. The ability to utilize **additive manufacturing with asteroid metals** in space would further enhance this capability, allowing for the creation of customized parts and tools on demand, reducing the need to pre-ship every component from Earth. Connecting this to the broader space economy, asteroid-derived metals represent an "enabling infrastructure" output, facilitating further development and expansion beyond Earth.

Water and Propellant for Deep Space Missions are another crucial output. The abundance of water ice in C-type asteroids offers a readily accessible source for **providing water ice** (NASA, 2005; NASA

2025) for life support on long-duration deep space missions to Mars, Europa, and other destinations in the outer solar system. This would significantly reduce the amount of water that needs to be transported from Earth for these ambitious voyages. More importantly, the ability to process this water ice into **propellant** – liquid hydrogen and oxygen – in space will be transformative. Establishing **in-space refueling depots** at strategic locations within the asteroid belt or at Lagrange points, supplied with asteroid-derived propellant, would dramatically reduce the propellant mass required for deep space missions to be launched from Earth, increasing payload capacity and making previously infeasible missions a reality. This capability represents a "fuel and sustenance" output, directly enabling more ambitious and far-reaching space exploration endeavors.

In the long-term, the **return of rare and valuable materials to Earth** represents another potential output of the asteroid belt supply chain, although this remains a vision with significant economic and technological uncertainties. The potential for extracting and returning **Rare Earth Elements**, which are critical for many high-tech industries on Earth, could provide a new and potentially abundant source for these increasingly sought-after materials. Similarly, the high value of Platinum Group Metals found in some asteroids could potentially disrupt terrestrial supply chains for these precious metals, which are used in a wide range of applications. However, the **challenges of return transport** – the immense energy requirements and long transit times – are substantial. The **economic uncertainties** surrounding the cost of extraction, processing, and transportation compared to terrestrial sources, as well as the volatility of commodity markets, need to be carefully considered. Furthermore, the potential for **market disruption** caused by a large influx of asteroid-derived precious metals would need to be managed. While the return of rare materials to Earth remains a long-term and complex prospect, it represents a potential "economic return" output of a mature asteroid belt supply chain.

Challenges and Ethical Considerations for Asteroid Belt Supply Chains

Establishing a thriving supply chain in the asteroid belt presents a multitude of immense challenges and raises important ethical considerations that must be addressed. The **vast distances and long mission durations** inherent in operating within the asteroid belt demand the development of **highly autonomous systems**. AI driven spacecraft and robotic miners will need to operate for years, or even decades, with minimal human intervention, capable of navigating, making decisions, and performing complex tasks independently. The significant communication delays between Earth and the asteroid belt, which can range from minutes to tens of minutes depending on the distance, preclude real-time remote control for many operations, further emphasizing the need for robust autonomy. Ensuring the **reliability and redundancy** of these systems is paramount, as failures in critical components could lead to mission loss or significant delays in resource production.

The **technological development for mining and processing in space** requires significant breakthroughs. The technologies needed for autonomously mining asteroids, efficiently processing and refining resources in the unique conditions of microgravity and vacuum, and transporting these materials over vast distances are still largely in the research and development phase. **Developing efficient and scalable techniques** that are also cost-effective will be crucial for the economic viability of asteroid mining. Rigorous **testing and validation in space** will be essential to ensure the functionality and reliability of these technologies before large-scale operations can commence.

The **economic viability and investment** required for asteroid mining are shrouded by unknown future technology developments. Accurately predicting the **economic return** on investment is challenging given the long development timelines, the uncertainties surrounding resource availability and market prices, and the high upfront costs associated with developing the necessary technologies and infrastructure. **Long-term investment** over decades will be required to bring the vision of an asteroid belt supply chain to fruition. **Attracting private investment** will depend on establishing clear regulatory frameworks, demonstrating the potential for profit, and mitigating the inherent risks associated with space activities.

Planetary protection and environmental concerns must also be carefully considered. While the asteroid belt is not known to harbor life as we understand it, the potential **environmental impact of large-scale mining operations** on the dynamics and stability of the asteroid belt needs to be understood and mitigated. Developing **responsible**

resource utilization practices that minimize disruption and ensure the long-term sustainability of this resource is crucial. The **potential for cross-contamination** between Earth and asteroids, while seemingly low, should also be considered and addressed through appropriate protocols.

Finally, the development of **space law and resource rights** is essential for governing the utilization of asteroid resources. The current lack of clear international legal frameworks governing resource extraction in space creates uncertainty and could lead to disputes. The ongoing **debate over resource ownership and rights** – who has the right to own and exploit resources extracted from asteroids – needs to be resolved through international dialogue and agreements. Establishing clear and equitable **international agreements** will be crucial for ensuring the responsible and sustainable development of the asteroid belt as a vital resource for the future of space exploration and development.

PART III: Cross-Cutting Themes and Future Directions

We have completed our journey across the solar system. The lessons learned and technology developed from one stage to another act as building blocks. Successes being built upon and failures becoming scar tissue. Once the initial exploration is complete, new civilizations will emerge and societies will form. The doors will be wide open for how they organize, where they establish themselves and what they see as their purposes for existence. Bold ideas and exciting experiments will spring up across multiple locations in our solar system. They will open the doors to a new stage in human societal evolution. Whereever we go and whatever we build one thing will remain unchanged, the basic needs of life.

Figure 9. Author's depiction of overlapping major themes of improved engines, legal domains, automated systems and closed loop living facilities.

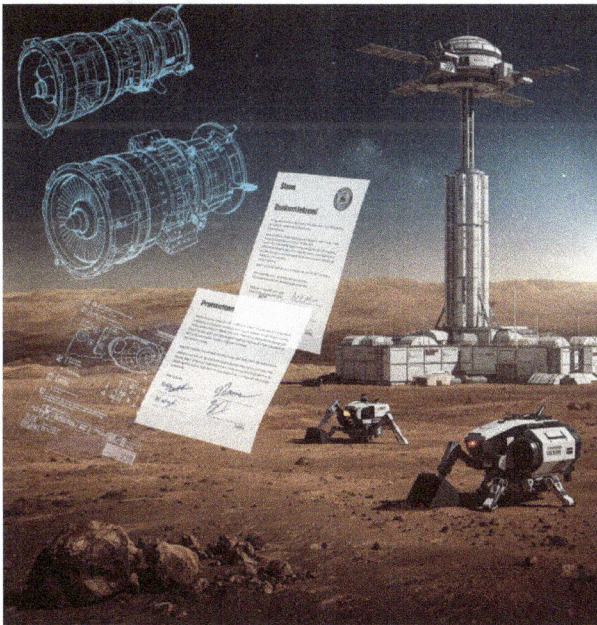

Chapter Six: Enabling Technologies and Infrastructure for Space-Based Supply Chains

The preceding chapters have laid out the compelling rationale for establishing supply chains beyond Earth, detailing the specific needs and potential of lunar, Martian, and asteroid belt operations. The realization of these ambitious endeavors hinges upon the development and deployment of a suite of **enabling technologies and robust infrastructure**. The most likely scenario will be development in stages with lessons learned on the Moon informing the planning and colonization of Mars, which in turn will inform expansion into the asteroid belt. This chapter will delve into the critical advancements in various fields that will be necessary to forge reliable pathways for the flow of resources and personnel throughout the solar system, directly impacting the feasibility, efficiency, and sustainability of our extraterrestrial supply chains.

Advanced Propulsion Systems

The very foundation of any space-based supply chain rests upon the ability to efficiently transport goods and people across vast distances. Traditional chemical propulsion, while foundational to our current near Earth demands, presents limitations in terms of speed, payload capacity, and overall mission cost for long-duration interplanetary travel. The future of space-based supply chains will be heavily reliant on the maturation and implementation of more advanced propulsion systems.

Nuclear Thermal Propulsion (NTP)

NTP systems offer the potential for significantly higher specific impulse (a measure of propellant efficiency) compared to chemical rockets. By heating a propellant (typically hydrogen) to extremely high temperatures using a nuclear reactor and expelling it through a nozzle, NTP can achieve thrust with less propellant, leading to faster transit times and greater payload capacity. Within the context of supply chains, NTP could revolutionize interplanetary transportation, drastically reducing the months-long journeys to Mars, as discussed in Chapter Four. This speedier transit would translate to faster resupply missions for Martian bases, quicker deployment of infrastructure, and more efficient return journeys, ultimately streamlining the entire Martian supply chain. Furthermore, NTP could make the exploration and utilization of resources in the asteroid belt, as detailed in Chapter Five, more feasible by reducing the travel time for robotic mining fleets and the transport of extracted materials to lunar or Earth orbits.

Advanced Chemical Rockets

While NTP offers a significant leap in efficiency, advancements in traditional chemical propulsion continue to play a vital role. Research into higher-performance propellants, such as liquid oxygen and methane (as considered for Martian ascent vehicles in Chapter Four), and improved engine designs with higher thrust-to-weight ratios will enhance the capabilities of chemical rockets. These advancements are crucial for specific segments of the supply chain, such as lunar landers (Chapter Three) requiring high thrust for landing and ascent, and potentially as powerful boosters for the initial stages of interplanetary transfer vehicles. Improved chemical propulsion will also be essential for efficient cargo transfer within planetary systems, for example, moving resources from Martian orbit to the surface.

Solar Electric Propulsion (SEP)

SEP systems utilize electric fields to accelerate ions, generating a very low but continuous thrust over long periods. While transit times are longer compared to chemical or nuclear propulsion, SEP offers exceptionally high propellant efficiency, making it ideal for transporting large cargo masses over interplanetary distances. Within the supply chain framework, SEP is particularly well-suited for the slow but steady transport of bulk resources from the asteroid belt (Chapter Five) to lunar or Martian orbit, or even back to Earth orbit. Fleets of SEP-powered cargo ships could continuously ferry metals, water ice, and other valuable materials, establishing a reliable and cost-effective logistical link across vast distances. Once the chain of consistent shipping is in place this offers the most efficient option for steady supplies of goods from the asteroid belt to Mars and Earth.

Beamed Energy Propulsion

This innovative concept involves using ground-based or space-based lasers or microwaves to beam energy to a spacecraft, which then uses this energy to heat a propellant or directly generate thrust. Beamed energy propulsion could offer a cost-effective way to launch small payloads from planetary surfaces, potentially supplementing traditional rocket launches for resupply missions to lunar or Martian outposts. It could also be used for in-space transportation of smaller, time-insensitive cargo within the supply chain network.

Solar Sail Propulsion

Solar sails offer a revolutionary approach to space propulsion by harnessing the momentum of photons from the Sun, providing a significant benefit in that they require no onboard propellant, drastically reducing mission mass and enabling potentially limitless operational durations. This fuel-less propulsion allows for continuous, albeit very gentle, thrust, which over extended periods can accelerate spacecraft to remarkably high speeds, making them ideal for long-duration interplanetary voyages, as discussed in Chapter Six. Furthermore, solar sails are a fundamentally clean and sustainable technology, producing no exhaust or pollution, aligning with the ethical considerations for space industrialization outlined in Chapter Seven. However, the very nature of solar sails also presents disadvantages; the thrust generated by sunlight is extremely weak, resulting in very slow acceleration and making them unsuitable for missions requiring rapid changes in velocity or high thrust maneuvers. Additionally, to capture sufficient sunlight, solar sails must be enormously large, posing significant engineering challenges in terms of deployment, structural integrity, and control in the harsh space environment, potentially increasing the complexity and up-front cost of the overall system.

Fusion Propulsion (Long-Term)

Fusion propulsion, while still in the realm of long-term development, holds the potential for extremely high specific impulse and thrust. Harnessing the power of nuclear fusion could revolutionize space travel, enabling rapid transit times throughout the solar system. In the context of supply chains, fusion propulsion could facilitate the fast and efficient transport of large quantities of resources and personnel between Earth, the Moon, Mars, and the asteroid belt, creating a truly interconnected and responsive solar system-wide logistical network. This capability will impact travel across the solar system in a s similar fashion to how aircraft revolutionized global travel.

In-Space Manufacturing and Robotics

To minimize reliance on Earth-launched supplies and to effectively utilize extraterrestrial resources, advanced in-space manufacturing and robotics capabilities are paramount. These technologies will enable the creation of tools, infrastructure, and even spacecraft components directly in the space environment.

3D Printing (Additive Manufacturing)

3D printing offers a transformative capability for space-based supply chains. By allowing for the on-demand production of tools, spare parts, habitat components, and even customized scientific equipment using digital designs, 3D printing significantly reduces the need to pre-ship a vast inventory of items from Earth. This is particularly crucial for establishing self-sufficient bases on the Moon and Mars (Chapters Three and Four) and for supporting long-duration robotic operations in the asteroid belt (Chapter Five). Imagine lunar colonists printing replacement parts for rovers using locally sourced regolith-based materials, or Martian settlers fabricating customized habitat modules using imported polymers and in-situ extracted resources. In the asteroid belt, robotic miners could even print replacement parts for themselves, ensuring continuous operation with minimal need for Earth intervention.

Autonomous Robotics

Autonomous robots will be the workhorses of space-based supply chains. From prospecting for resources on asteroids (Chapter Five) and extracting water ice on the Moon and Mars (Chapters Three and Four) to assembling large structures in orbit and maintaining infrastructure on planetary surfaces, robots capable of operating with minimal human intervention are essential. Advanced sensors, artificial intelligence, and robust control systems will be required to enable these robots to perform complex tasks in harsh environments, navigate challenging terrain, and coordinate their efforts across vast distances. In the asteroid belt, fleets of autonomous robots will be critical for the economic viability of resource extraction, operating independently for extended periods to locate, mine, and process valuable materials.

Advanced Materials

The development and utilization of advanced materials will be crucial for building durable and efficient infrastructure within space-based supply chains. Materials with high strength-to-weight ratios will be essential for constructing lightweight spacecraft and habitats, reducing launch costs from Earth. Radiation-resistant materials will be vital for protecting astronauts and sensitive equipment from the harsh space environment on the Moon, Mars, and during interplanetary transit. Self-healing materials could extend the lifespan of infrastructure and reduce the need for frequent repairs. The ability to manufacture these advanced materials in space, potentially using asteroid-derived resources, would further enhance the self-sufficiency of the supply chain.

AI for In-Space Operations

Artificial intelligence will play an increasingly important role in managing and optimizing space-based supply chains. AI algorithms can be used to analyze vast amounts of data from prospecting missions, optimize mining operations, plan efficient transportation routes, predict equipment failures, and manage complex logistical networks across multiple celestial bodies. In the asteroid belt, AI will be critical for coordinating the activities of large fleets of autonomous robots, making real-time decisions in response to changing conditions, and ensuring the efficient flow of resources to processing facilities and transport vehicles.

Modular Construction

Designing habitats, spacecraft, and infrastructure using modular components offers significant advantages for space-based supply chains. Modular designs allow for easier transportation of individual units, simplified assembly in the microgravity environment of space, and greater flexibility for expansion and repair. Imagine lunar or Martian bases constructed from interconnected, standardized modules that can be launched separately and assembled on the surface. Similarly, spacecraft could be built from modular components, allowing for easier upgrades and customization. This modular approach simplifies the logistics of transporting and deploying large structures within the supply chain network.

Closed-Loop Life Support Systems

For long-duration missions and permanent settlements beyond Earth, minimizing the reliance on resupply from our home planet is crucial. Closed-loop life support systems (CLSS) are a cornerstone of sustainable human presence beyond Earth. These intricate systems are designed to operate like micro ecosystems, mimicking the Earth's natural processes of recycling and resource utilization. Just as in a terrestrial ecosystem where plants convert sunlight into energy, consume carbon dioxide, and release oxygen, while animals consume plants and release carbon dioxide, a CLSS seeks to create a self-sustaining balance between the needs of the human crew and the regeneration of essential resources Closed-loop life support systems, which recycle essential resources, will be a cornerstone of sustainable space-based supply chains.

Advanced Recycling

Implementing advanced recycling technologies for water, air, and waste within space habitats will significantly reduce the need for costly and frequent resupply missions from Earth. This includes sophisticated water purification systems that can reclaim water from wastewater and even humidity in the air, as well as air revitalization systems that remove carbon dioxide and replenish oxygen. Efficient waste management systems that can process and potentially recycle waste products will also be essential. For long-duration missions to Mars (Chapter Four) and permanent settlements on the Moon, Mars, and in orbital habitats, these advanced recycling systems will be critical for the sustainability of the supply chain.

Bioregenerative Systems

Taking resource recycling a step further, bioregenerative life support systems integrate biological components, such as algae and plants, into the life support loop. These systems can produce food for the crew, generate oxygen through photosynthesis, and help to purify water and air, mimicking Earth's natural ecosystems. While still under development, bioregenerative systems hold immense potential for achieving greater self-sufficiency in long-term space settlements, particularly on Mars where the potential for in-situ agriculture is being explored (Chapter Four). By producing food and oxygen locally, these systems significantly reduce the logistical burden on the supply chain from Earth.

Minimizing Resource Consumption

Beyond recycling, designing systems and operations to minimize the overall consumption of resources is crucial. This includes developing energy-efficient technologies, implementing water conservation measures, and optimizing the use of consumables. Every kilogram of resource saved in space translates to a kilogram less that needs to be launched from Earth, directly impacting the cost and efficiency of the supply chain.

Energy Systems for Space Operations

A reliable and abundant supply of energy is fundamental to powering all aspects of space-based supply chains, from life support systems in habitats to resource extraction and processing facilities.

Advanced Solar Power

Solar power will continue to be a primary energy source for many space operations. Advancements in solar panel technology, such as higher efficiency, lighter weight, and the development of large, deployable structures, will allow for the generation of significant amounts of power in space. This energy can be used to power habitats, rovers on the Moon and Mars, and even SEP systems for transporting cargo within the asteroid belt. However, the limitations of solar power, such as the lack of sunlight during lunar nights (Chapter Three) and the reduced intensity of sunlight further from the Sun, necessitate the development of complementary energy sources.

Nuclear Power (Fission/Fusion)

Nuclear power offers a reliable and high-power energy source that is not dependent on sunlight. Small, modular nuclear fission reactors could provide continuous power for lunar bases (Chapter Three) and Martian settlements (Chapter Four), enabling resource extraction, in-situ manufacturing, and life support systems. In the longer term, fusion power, with its potential for even greater energy output, could power large-scale industrial operations in space and provide the energy needed for advanced propulsion systems. Ensuring the safe deployment and operation of nuclear power systems in space, as well as addressing the logistical considerations for their transport and maintenance, will be crucial.

Beamed Energy

As mentioned earlier, beamed energy can also be used to provide power to remote locations in space. Ground-based or space-based lasers or microwaves could beam energy to receivers on lunar or Martian rovers, providing them with a continuous power supply even during the long Martian dust storms or lunar nights. This could enhance the operational capabilities of robotic assets within the supply chain. A variation of this is space based solar collectors that beam energy to other space platforms or surface collectors.

Micro Supply Chains

Within a CLSS, a miniature supply chain emerges. Water, air, and even certain food waste are continuously recycled and reused. This involves a series of interconnected subsystems, each playing a vital role in maintaining the delicate balance. Water purification systems remove impurities and contaminants from wastewater, making it suitable for drinking and hygiene. Air revitalization systems remove carbon dioxide and other impurities from the atmosphere, while simultaneously generating oxygen through processes like electrolysis or by utilizing photosynthetic organisms. Waste management systems break down organic waste and recycle inorganic materials, minimizing the accumulation of waste within the habitat. These subsystems, with their inputs, outputs, and interconnectedness, can be viewed as a microcosm of the larger space-based supply chains discussed throughout this book.

Lunar, Martian, and Asteroid Applications

While the fundamental principles of CLSS remain the same across different locations, the specific design and implementation will vary depending on the unique challenges of each environment.

Lunar CLSS: On the Moon, the primary challenge will be managing resource scarcity. Water ice, a potential resource identified in Chapter Three, will be crucial for both life support and propellant production. Lunar dust presents unique challenges, as it can be abrasive and potentially contaminate systems. The lunar environment also lacks a significant atmosphere, requiring robust radiation shielding and protection from micrometeoroid impacts.

Martian CLSS: Mars presents a more complex environment. The Martian atmosphere, though thin, offers a potential source of CO_2 for oxygen production, as discussed in Chapter Four. However, the presence of perchlorates in the Martian soil (Rymski, et al., 2024) requires careful consideration in designing life support systems and agricultural processes. The Martian environment also offers the potential for in-situ resource utilization, with the possibility of cultivating certain crops using Martian soil and atmospheric resources.

Asteroid-Based CLSS (for long-duration missions): While asteroid mining operations will likely be initially automated, the long-term vision may include human presence in the asteroid belt. In this scenario, CLSS will need to be highly robust and capable of operating independently for extended periods with minimal resupply from Earth. The focus will be on maximizing resource utilization and minimizing waste generation, as the distance to Earth will make resupply missions extremely challenging and expensive.

Closed-loop life support systems are not just technological marvels; they represent a required fundamental shift in our approach to human exploration. By creating self-sustaining micro-ecosystems within our space habitats, we can reduce our reliance on Earth-based resupply, enhance the long-term viability of off-world settlements, and pave the way for a truly sustainable human presence in the cosmos

Communication and Navigation Infrastructure

Maintaining constant communication and ensuring precise navigation are essential for the safe and efficient operation of space-based supply chains. Being able to communicate between humans will also be vital for both social and business reasons. The vast distances between parts of our solar system create many challenges to reliable and rapid communication.

Deep Space Communication Networks

Robust and high-bandwidth communication networks are vital for transmitting data, sending commands, and ensuring the safety of spacecraft and personnel operating throughout the solar system. Existing networks like NASA's Deep Space Network and ESA's

Estrack will need to be expanded and enhanced to support the increasing demands of a growing space-based economy. The development of new communication technologies, such as laser communications, will also be crucial for increasing data transfer rates over vast distances. This infrastructure will be essential for managing robotic fleets in the asteroid belt (Chapter Five) and for maintaining contact with crewed missions to Mars (Chapter Four).

Advanced Navigation Systems

Precise navigation is critical for all aspects of space-based supply chains, from accurately landing on the Moon or Mars to rendezvous and docking with asteroids or orbital stations. This will require the development of advanced navigation systems, potentially including GPS-like networks on the Moon and Mars to provide localized positioning data, as well as highly accurate inertial measurement units for spacecraft navigating interplanetary space. Just like ancient travelers on Earth, the sun remains an excellent orientation point of reference for solar positioning. Autonomous navigation capabilities, relying on sophisticated sensors and onboard processing, will be particularly important for missions to the distant asteroid belt.

Autonomous Operation Capabilities

Reducing the reliance on constant communication with Earth, especially for operations in the outer solar system, necessitates the development of highly autonomous spacecraft and robotic systems. These systems will need to be capable of making decisions, adapting to changing conditions, and troubleshooting problems independently, relying on onboard processing power and sophisticated software. This is particularly crucial for the long-duration missions and vast distances involved in asteroid belt resource utilization (Chapter Five).

Spaceports and Orbital Infrastructure

Just as terrestrial supply chains rely on ports, airports, and transportation hubs, space-based supply chains will require a network of spaceports and orbital infrastructure to facilitate the movement of goods and people. These will also develop progressively from orbital Earth waypoints in Lagrange positions, to Lunar based hubs out to Martian ports and eventually in the asteroid belt. The self-sufficiency requirements will grow the farther from Earth the ports operate.

Spaceports on Earth, Moon, Mars, and potentially in Orbit

Traditional Earth-based spaceports will continue to play a vital role in launching the initial components and personnel for space-based supply chains. However, as we establish a more permanent presence beyond Earth, the development of spaceports on the Moon (Chapter Three) and Mars (Chapter Four) will become essential for facilitating surface-to-orbit and interplanetary travel of locally sourced resources and personnel. The concept of orbital spaceports, located at strategic points in Earth orbit, lunar orbit, or Martian orbit, could serve as staging points for deep space missions, assembly hubs for large spacecraft, and transfer nodes within the supply chain network.

Orbital Transfer Vehicles (OTVs)

Dedicated Orbital Transfer Vehicles will be crucial for moving payloads and potentially crew between different orbits around Earth, the Moon, and Mars. These reusable spacecraft, equipped with efficient propulsion systems, will act as the workhorses of the orbital supply chain, delivering resources to specific locations, assembling large structures, and ferrying personnel between orbital stations and planetary surfaces.

Space Tugs

Smaller, more agile "space tugs" will be needed for maneuvering payloads in orbit, servicing spacecraft, and potentially retrieving smaller resources from near-Earth asteroids or within the asteroid belt. These versatile vehicles will play a vital role in the fine-grained logistics of the space-based supply chain.

Fuel Depots

As highlighted in previous chapters, fuel depots located at strategic points in Earth orbit, lunar orbit, Martian orbit, and potentially within the asteroid belt will be critical for enabling efficient and cost-effective space transportation. These depots will allow spacecraft to refuel in space, reducing the amount of propellant needed to be launched from planetary surfaces and significantly increasing payload capacity and mission range. Establishing the logistics for supplying these fuel depots with both Earth-derived and in-situ produced propellant will be a key aspect of building a sustainable space-based supply chain.

Cybersecurity and Space Domain Awareness

As our reliance on space-based infrastructure grows, ensuring the security and safety of these assets becomes paramount. Competition for resources and preferred locations for bases are already in motion. Cyberattacks are common across our planet and there is no reason to believe they will do anything but expand along with a growth in space.

Protecting Space-Based Supply Chains from Threats

The interconnected network of spacecraft, ground control systems, and data links that constitute a space-based supply chain will be vulnerable to cybersecurity threats. Protecting this critical infrastructure from hacking, data breaches, and other malicious activities will be essential to prevent disruptions to the flow of resources and information. Robust cybersecurity measures, including encryption, intrusion detection systems, and secure communication protocols, will need to be implemented across the entire supply chain network.

Ensuring Safe and Responsible Operations

With an increasing number of actors and assets operating in space, maintaining Space Domain Awareness (SDA) – the ability to track and identify objects in orbit – is crucial for ensuring the safe and responsible operation of space-based supply chains. Accurate tracking of spacecraft, debris, and other objects will help to prevent collisions and ensure the long-term sustainability of space activities. International cooperation and the development of clear regulations for space traffic management will be essential for promoting responsible behavior and protecting the space environment.

The establishment of robust and sustainable space-based supply chains is an ambitious undertaking that will require significant advancements and the seamless integration of numerous enabling technologies and infrastructure elements. From advanced propulsion systems that will shrink the distances of the solar system to in-space manufacturing and robotics that will empower self-sufficiency, and from closed-loop life support systems that will sustain long-duration missions to the critical communication, navigation, and energy infrastructure that will tie it all together, the technologies discussed in this chapter represent the essential building blocks for forging the pathways beyond Earth and unlocking the vast potential of our cosmic neighborhood. The development and deployment of these capabilities, coupled with a commitment to cybersecurity and responsible operations, will pave the way for a future where humanity is a truly spacefaring species, sustained by the intricate and innovative supply chains that extend throughout the solar system.

Chapter Seven: Economic, Policy, and Strategic Considerations for Space-Based Supply Chains

As our abilities to survive and shape the living spaces of our new homes mature, the organizational features gain prominence. Having meticulously detailed the technological and infrastructural requirements for establishing supply chains beyond Earth, we need to explore the unique challenges and opportunities presented by the Moon, Mars, and the asteroid belt. However, the realization and long-term viability of these ambitious extraterrestrial logistical networks are not solely dependent on engineering prowess. A robust framework encompassing economic viability, supportive policies, and strategic considerations is equally crucial to ensure their success and sustainability. This chapter will delve into these essential non-technical aspects, demonstrating how they directly influence the development and operation of space-based supply chains.

Figure 10: Author's rendition of board meetings in orbit.

Economic Models and Business Cases

The establishment of space-based supply chains represents a significant undertaking with substantial upfront investment. Therefore, a thorough understanding of the underlying economic models and the development of compelling business cases are paramount. Analyzing the **economics of space-based supply chains** involves considering the high initial costs associated with research, development, and deployment of space technologies and infrastructure, as highlighted in Chapter Six. This includes the cost of advanced propulsion systems for transporting resources (as discussed in Chapters Three, Four, and Five), the expense of developing autonomous mining and processing facilities for lunar regolith, Martian water ice, and asteroid metals, and the investment required for establishing orbital depots and surface bases. However, a comprehensive economic analysis must also account for the potential for significant long-term cost savings derived from utilizing in-space resources. As emphasized in Chapters Three, Four, and Five, the ability to produce water, propellant, and construction materials on the Moon, Mars, and from asteroids can drastically reduce the reliance on expensive and energy-intensive launches from Earth, ultimately making space operations more sustainable. Furthermore, the economics of scale will play a crucial role as the volume of resources transported and utilized within these supply chains increases over time. Efficient transportation models, such as reusable rockets (as pioneered by SpaceX and discussed in Chapter Six) and Solar Electric Propulsion tugs for bulk cargo transport (also in Chapter Six), will be key to optimizing the cost-effectiveness of the entire supply chain.

The development of viable **potential revenue streams** will be essential to attract investment and ensure the long-term financial sustainability of space-based supply chains. As explored in Chapter Five, the asteroid belt holds immense potential for providing valuable resources like water and propellant for deep space missions, creating a market for in-space refueling services. The metals extracted from asteroids can also be sold for in-space construction and manufacturing, as discussed in Chapter Five, reducing the need to launch these materials from Earth for building habitats and infrastructure in lunar or Martian orbit. The long-term vision of returning rare and valuable materials from asteroids to Earth (Chapter Five) represents another potential revenue stream, although its economic viability is still under evaluation. Closer to home, the utilization of lunar resources, such as water ice for propellant (Chapter Three), and Martian resources, potentially including locally produced food and construction materials (Chapter Four), can support local operations and potentially even facilitate off-world trade within the broader solar system supply chain. Moreover, the enhanced efficiency and frequency of scientific research and exploration enabled by robust supply chains will also have an indirect economic value, driving innovation and potentially leading to new discoveries with terrestrial applications.

Attracting the necessary **investment** for these ambitious undertakings will require a multifaceted approach. **Government funding** will likely play a crucial role in the initial stages, supporting foundational research and development, as well as establishing essential infrastructure. **Private investment** and **venture capital** will be attracted by the potential for long-term profitability and the development of new markets in space. Given the long development timelines and the inherent risks associated with space activities, **long-term investment horizons** and the availability of **patient capital** will be critical. A **phased investment** strategy, starting with lower-risk and potentially earlier-return activities on the Moon (Chapter Three) and gradually expanding to the more complex and long-term endeavors on Mars and in the asteroid belt (Chapters Four and Five), may be a pragmatic approach.

Public-private partnerships offer a promising model for sharing the risks and rewards of developing space-based supply chains. By leveraging the strengths of both the public and private sectors, these collaborations can accelerate innovation and ensure the efficient operation of logistical networks. Successful models of public-private collaboration in the space industry, such as NASA's Commercial Crew Program, can provide valuable lessons for establishing partnerships focused on resource utilization and in-space logistics. Governments can play a vital role in setting strategic goals, providing initial funding, and establishing regulatory frameworks, while private companies can bring their expertise in technology development, operational efficiency, and market creation to the table.

Space Policy and Regulatory Frameworks need to be developed. The effective iterations and operation of space-based supply chains will necessitate a clear and supportive policy and regulatory environment at both the national and international levels. The existing framework of **international space law**, primarily based on the Outer Space Treaty of 1967, provides a foundation but also presents ambiguities regarding the utilization of space resources. While the treaty prohibits national appropriation of celestial bodies, its stance on the extraction and ownership of resources is less clear. This ambiguity, as touched upon in Chapter Five in the context of asteroid mining, creates uncertainty for potential investors and could hinder the development of a robust legal environment for space-based supply chains. Evolving **international agreements** will likely be necessary to provide a clear and stable legal framework that balances the interests of all nations and promotes the responsible utilization of space resources.

The debate surrounding **resource rights** – who owns the resources extracted from the Moon, Mars, and asteroids – is a critical policy consideration. Different perspectives exist, ranging from the "common heritage of mankind" principle, which suggests that space resources should benefit all humanity, to the view that private entities should have the right to own and exploit the resources they extract. National legislation, such as the US Commercial Space Launch Competitiveness Act, which grants US citizens the right to own asteroid resources they mine, reflects a move towards the latter perspective. However, a lack of international consensus on this issue could lead to disputes and hinder the development of a globally recognized and functional supply chain.

Environmental regulations will be essential to govern the activities associated with space-based supply chains. This includes establishing guidelines for responsible mining practices on the Moon, Mars, and asteroids (as mentioned in Chapter Five) to minimize environmental disruption and protect potentially pristine environments. Regulations will also be needed for in-space processing activities to prevent pollution and ensure the safe handling of potentially hazardous materials. Furthermore, as highlighted in Chapter Six, the increasing volume of space traffic associated with supply chain operations will necessitate effective **space traffic management** systems to ensure the safe and efficient movement of spacecraft and resources, preventing collisions and mitigating the creation of space debris. National space agencies and international organizations will play crucial roles in developing and enforcing these environmental regulations and space traffic management rules.

Governments and international organizations have a fundamental role to play in shaping the policy and regulatory landscape for space-based supply chains. Governments can set strategic goals for space exploration and resource utilization, provide crucial funding for foundational research and development, and establish national regulatory frameworks that encourage responsible private sector involvement. International organizations, such as the United Nations, can facilitate the development of international space law, promote cooperation among nations in the utilization of space resources, and help to establish global standards for environmental protection and space traffic management.

Strategic Implications and Geopolitics of Space Resources: The development of space-based supply chains has profound strategic implications and will undoubtedly influence the geopolitics of the future.

The potential for **competition for space resources** among nations and private companies is a significant strategic consideration. As highlighted in Chapters Three, Four, and Five, resources like water ice (essential for propellant and life support), rare earth elements, and platinum group metals hold significant economic and strategic value. The control over access to and the means of extracting and processing these resources could become a source of tension and potentially reshape international relations, leading to new geopolitical dynamics in space.

National security considerations are also intertwined with the development of space-based supply chains. Resource independence, particularly for critical materials needed for space infrastructure and defense technologies, could become a key national security objective.

The development of dual-use technologies, applicable to both commercial and military space activities, is also likely. Furthermore, the control over strategic transportation routes in space and key resource nodes on the Moon, Mars, and in the asteroid belt could become strategically important for maintaining national power and influence in the space domain. Military space programs will likely play a role in protecting space assets and ensuring access to space for national interests.

The development of space-based supply chains presents opportunities for both **international cooperation and competition**. While competition for resources may arise, there is also significant potential for nations to collaborate on research, technology development, and the establishment of shared infrastructure. International cooperation could help to share the costs and risks associated with these ambitious endeavors and promote a more stable and predictable environment for space resource utilization. However, managing competing national interests and ensuring equitable access to resources will be a significant challenge. Historical examples of international cooperation in space, such as the International Space Station, offer valuable lessons that can be applied to the development of space resource utilization frameworks.

Ultimately, access to and control over space resources and the infrastructure of space-based supply chains could become a significant factor in **global power dynamics** in the future. Nations and entities that can effectively leverage these resources and logistical networks will likely gain economic and strategic advantages, potentially shifting the balance of power on Earth and shaping the future of human civilization in space.

Sustainability and Ethical Considerations: As we venture beyond Earth and begin to utilize extraterrestrial resources, it is imperative that we do so in a sustainable and ethical manner.

Responsible resource utilization is crucial to ensure the long-term viability of space-based supply chains and to avoid the depletion of valuable resources. This will require the development of resource management strategies, including efficient extraction and processing techniques, as well as the potential for in-situ recycling and reuse of materials to extend their lifespan and minimize waste. The environmental concerns associated with asteroid mining, as mentioned in Chapter Five, highlight the need for careful planning and the adoption of best practices to minimize disruption to these pristine environments.

Planetary protection is another critical ethical and scientific consideration. Preventing the contamination of other celestial bodies, such as Mars or Europa, with Earth-based microbes is essential to safeguard potential future scientific discoveries related to extraterrestrial life. Similarly, precautions must be taken to prevent the introduction of potentially harmful extraterrestrial materials to Earth. Robust protocols and procedures will need to be implemented to ensure that resource extraction and other activities in space are conducted in a manner that minimizes the risk of biological contamination.

The growing problem of **space debris** in Earth orbit, as highlighted in Chapter Six, serves as a cautionary tale for future space activities. As the scale of space-based supply chain operations increases, it will be crucial to minimize the creation of new space debris in all orbital regimes. This will require responsible spacecraft design, the implementation of effective end-of-life disposal strategies for spacecraft and infrastructure, and potentially the development of active debris removal technologies. Maintaining a safe and sustainable space environment is essential for the long-term success of space-based supply chains.

Finally, the large-scale industrialization of space and the potential for human expansion beyond Earth raise profound **ethical implications**. Questions related to the moral status of extraterrestrial environments, the potential for exploitation of resources, and the rights and responsibilities of humans living and working off-world will need to be carefully considered as we move towards a future where space-based supply chains play an increasingly significant role in human civilization.

Workforce Development and Education will take on both old and new paradigms. The development and operation of space-based supply chains will require a highly skilled and knowledgeable workforce. Building the **skilled workforce needed** for this endeavor will require a concerted effort across various disciplines. This includes engineers specializing in propulsion, robotics, materials science, and aerospace engineering, as well as experts in mining, in-space processing, logistics, and even space law and policy. Identifying potential gaps in the current workforce and proactively addressing these needs through targeted education and training programs will be crucial.

Education and training programs at universities, vocational schools, and within the space industry itself will need to adapt to prepare the next generation workforce for the unique challenges of developing and operating space-based supply chains. This will likely involve interdisciplinary education that combines technical knowledge with a strong understanding of the economic, policy, and ethical considerations of space development. International collaboration in developing educational standards and sharing best practices can also play a vital role.

Attracting and retaining **talent** in the space industry will be essential for the long-term success of space-based supply chains. This will require offering competitive salaries and benefits, providing opportunities for challenging and meaningful work, and fostering a culture of innovation and growth. Government policies and incentives that support the space industry and encourage young people to pursue careers in related fields will also be important.

The realization of sustainable and beneficial space-based supply chains hinges not only on technological and infrastructural advancements but also on careful consideration of the underlying economic models, the establishment of supportive policy and regulatory frameworks, an understanding of the strategic implications and geopolitical landscape, a commitment to sustainability and ethical principles, and the development of a skilled and dedicated workforce.

Addressing these multifaceted considerations in a holistic and forward-thinking manner will be crucial for unlocking the full potential of our extraterrestrial resources and ensuring a prosperous and sustainable future for humanity in space, driven by the efficient and responsible operation of our interstellar supply chains.

Epilogue: Charting the Course for Cosmic Expansion Through Interstellar Logistics

As we reach the culmination of our exploration into the intricate world of space-based supply chains, it is clear that humanity stands at a pivotal moment. The journey from our current Earth-dependent space activities to a future where our presence extends throughout the solar system is not merely a matter of technological advancement, but fundamentally a logistical challenge. This book has sought to illuminate the path forward, detailing the resources, the needs, the enabling technologies, and the crucial economic, policy, and strategic considerations that will shape our interstellar supply networks.

Recap of Key Themes and Findings

Our journey began by envisioning a future where humanity's footprint extends beyond Earth, driven by the allure of scientific discovery, the promise of abundant resources, and the imperative for long-term survival. We then delved into the specific potential of the Moon, our nearest celestial neighbor, as a crucial initial node in a space-based supply chain (Chapter Three). Its proximity and resources like water ice offer a stepping stone for developing the foundational skills and infrastructure needed for more ambitious ventures. We then turned our attention to Mars (Chapter Four), a world with the potential to host a self-sustaining human civilization, but one that demands a far more complex and robust supply chain to overcome the challenges of distance and a more demanding environment. Finally, we looked towards the asteroid belt (Chapter Five), a vast reservoir of raw materials that could fuel a solar system-wide economy, provided we can develop the autonomous systems and transportation networks to access its riches. Underlying all these explorations has been the constant thread of the supply chain – the necessity of establishing reliable and efficient systems for sourcing, transporting, processing, and utilizing resources in the harsh and unforgiving vacuum of space.

Vision for a Multi-Planetary Future

The future enabled by robust space-based supply chains is one of unprecedented opportunity. Imagine lunar bases not just as flags planted on another world, but as thriving research outposts and resource processing centers, providing propellant and materials for missions further afield. Picture Martian settlements growing from small initial habitats into self-sufficient communities, contributing to scientific breakthroughs and perhaps even sending resources back to Earth. Envision the asteroid belt as a bustling hub of robotic activity, providing the raw materials needed to build vast space habitats, power deep space exploration, and potentially even alleviate resource scarcity on our home planet. This multi-planetary future, underpinned by efficient interstellar logistics, fosters scientific discovery on an unparalleled scale, drives economic growth through the creation of new industries and markets in space, and ultimately provides a more resilient future for humanity by diversifying our presence beyond a single world.

Remaining Challenges and Open Questions

While the potential is immense, the path to realizing this vision is fraught with significant challenges. As discussed in Chapter Six, substantial technological breakthroughs are still needed in areas like advanced propulsion, autonomous robotics, in-space manufacturing, and closed-loop life support. The economic viability of many aspects of space resource utilization remains uncertain, requiring innovative business models and sustained investment, as explored in Chapter Seven. Furthermore, the policy and regulatory frameworks governing space resource rights and environmental protection are still evolving and require international consensus (also in Chapter Seven). Ethical considerations surrounding our expansion into space and the potential impact on extraterrestrial environments must also be carefully addressed (Chapter Seven).

Call to Action

The realization of space-based supply chains is not a passive endeavor; it requires proactive engagement and a concerted effort from various stakeholders. Further research and development are crucial to mature the enabling technologies discussed in Chapter Six. Increased investment, both public and private, is needed to fund the ambitious projects required to establish the necessary infrastructure. International cooperation is paramount to navigate the complex legal, ethical, and strategic challenges outlined in Chapter Seven and to ensure that the benefits of space resource utilization are shared by all humanity.

Final Thoughts: The Transformative Potential of Space-Based Supply Chains

The establishment of robust space-based supply chains represents a transformative undertaking with the potential to redefine humanity's relationship with the cosmos. By taking the necessary action steps – by pushing the boundaries of technology, fostering economic growth in space, establishing clear and ethical policies, and working together on a global scale – we can unlock the vast potential of the solar system. This will not only benefit us through scientific discoveries and potential economic returns but will also provide a crucial insurance policy for our species, ensuring our long-term survival and enabling us to reach for the stars in ways previously unimaginable. The future of humanity lies beyond Earth, and the pathways to that future will be paved by the ingenuity and determination with which we build our interstellar supply chains.

Bibliography

Anderson, P (2023, September 21). There may be less water ice on the Moon than we thought. EarthSky. Retrieved from https://earthsky.org/space/water-ice-on-the-moon-permanently-shadowed-regions/

Berger, E. (2024, March 13). Mining helium-3 on the Moon has been talked about forever—now a company will try. Ars Technica. Retrieved from https://arstechnica.com/space/2024/03/mining-helium-3-on-the-moon-has-been-talked-about-forever-now-a-company-will-try/

Cordell, B. (1985). The Moons of Mars: A Source of Water for Lunar Bases and LEO. Lunar and Planetary Institute. Retrieved from https://adsabs.harvard.edu/full/1985lbsa.conf..809C#:~:text=Phobos%20and%20Deimos%20appear%20to,reservoir%20may%20be%2010%20grams.

Choudhury, R. (2023, September 16). Moon's ice not as old as believed, study finds. Interesting Engineering. Retrieved from https://interestingengineering.com/science/moons-ice-not-as-old-as-believed-study-finds?group=test_a

Fanale, F., Savale, J. (1989). Loss of water from Phobos. Geophysical Research Letters. 16(4), 287-290. DOI: 10.1029/GL016i004p00287.

Hu, R., Kass, D., Ehlmann, B., Yung, Y. Tracing the fate of carbon and the atmospheric evolution of Mars. Nature Communications, 2015; 6: 10003 DOI: 10.1038/NCOMMS10003

Kulu, E. (2022). In-space manufacturing-2022 industry survey and commercial landscape. In 20th IAA Symp on Building Blocks for Future Space Exploration and Development, Paris, France.

Kuthuner, S. (2023, September 20). Water ice on the moon may be key for future space missions. But is there enough? Space.com. Retrieved from https://www.space.com/future-moon-missions-find-less-water-than-expected-study

Margot, J. L., Brown, M. E. (2003) A Low-Density M-type Asteroid in the Main Belt. Science. 300: 1939-1942. doi:10.1126/science.1085844

Murray, J., Jagoutz, O. (2024). Olivine alteration and the loss of Mars' early atmospheric carbon. Science Advances, 2024; 10 (39) DOI: 10.1126/sciadv.adm8443

NASA. (2025). Asteroid Facts. NASA. Retrieved from https://science.nasa.gov/solar-system/asteroids/facts/.

NASA Hubble Mission Team. (2005, September 7). Largest Asteroid May Be 'Mini Planet' with Water Ice. NASA. Retrieved from https://science.nasa.gov/missions/hubble/largest-asteroid-may-be-mini-planet-with-water-ice/.

Penprase, B.E. (2023). Exploring the Earth and Sky. Models of Time and Space from Astrophysics and World Cultures. Astronomers' Universe. Springer, Cham. DOI: 10.1007/978-3-031-27890-7_1

Rzymski, R., Losiak, A., Heinz, J., Szukalska, M., Florek, E., Poniedziałek, B., Kaczmarek, L., Schulze-Makuch, D. (2024). Perchlorates on Mars: Occurrence and implications for putative life on the Red Planet. Icarus. 421. DOI: 10.1016/j.icarus.2024.116246.

Saeed, F., & Azeema, N. (2025). Role of Constellations in Shaping Cultural Identity across Civilizations, Celestial Myths about Constellations. The Critical Review of Social Sciences Studies, 3(1), 444-453. DOI: 10.59075/rpmtna03

Sanshua S. (2025 February 5). NASA finds the energy of the future on the Moon: There could be 3 million tons. Econews. Retrieved from https://www.ecoticias.com/en/nasa-finds-energy-on-the-moon/11091/

Schultz, C. (2014, April 8). Mars' Super-Thin Atmosphere May Mean that Flowing Water Was the Exception, Not the Rule. Smithsonian Magazine. Retrieved from https://www.smithsonianmag.com/smart-news/mars-super-thin-atmosphere-may-mean-flowing-water-was-exception-not-rule-180951158/

Tillman, N. (2017, May 4). Asteroid belt: Facts & formation. Space.com. Retrieved from https://www.space.com/16105-asteroid-belt.html.

Tomaswick, A. (2025, March 18). A new company plans to prospect the moon. Phys.org. Retrieved from https://phys.org/news/2025-03-company-prospect-moon.html#google_vignette

Waters, T., Campbell, B., Leuschen, C., Morgan, G., Cicchetti, A., Orosei, R., Plaut, J. (2024). Evidence of Ice-Rich Layered Deposits in the Medusae Fossae Formation of Mars. Advance Earth and Space Sciences. DOI: 10.129/2023GL105490.

Weiss, P. (2021). The Global Positioning System (GPS): Creating Satellite Beacons in Space, Engineers Transformed Daily Life on Earth. Engineering. 7 (3), 290-303. DOI: 10.1016/j.eng.2021.02.001.

ABOUT THE AUTHOR

Dr. and MAJ(R) David F. Bigelow, a retired member of the U.S. Army, resides in Thornton, Colorado with his wife. Throughout his career, Dr. Bigelow has held various roles within the Army and civilian sectors. His experience in aerospace spans both government and civilian positions. He has worked in Project Management, contracting, Supply Chain Team Lead and Director roles. While on active duty he managed contracts for Lockheed Martin Space Systems and United Launch Alliance. He has represented DCMA in the new assured access to space acquisition strategy IPT and supervised the Quality Assurance Team at ULA.

Dr. Bigelow's Army service includes tours of duty in several U.S. bases, Italy, Germany, Kuwait, Iraq, Afghanistan, and Kosovo. He has served as both an enlisted and officer infantryman. He is Airborne Ranger qualified and served on a variety of aerospace, logistics and combat missions.

www.ingramcontent.com/pod-product-compliance
Lightning Source LLC
Chambersburg PA
CBHW071557200326
41519CB00021BB/6797